Longjun Liu

Optimize Structural Topology and Computer Experimental Design

Longjun Liu

Optimize Structural Topology and Computer Experimental Design

with Simulation and Optimizers

VDM Verlag Dr. Müller

Impressum/Imprint (nur für Deutschland/ only for Germany)
Bibliografische Information der Deutschen Nationalbibliothek: Die Deutsche Nationalbibliothek
verzeichnet diese Publikation in der Deutschen Nationalbibliografie; detaillierte bibliografische
Daten sind im Internet über http://dnb.d-nb.de abrufbar.
Alle in diesem Buch genannten Marken und Produktnamen unterliegen warenzeichen-, marken-
oder patentrechtlichem Schutz bzw. sind Warenzeichen oder eingetragene Warenzeichen der
jeweiligen Inhaber. Die Wiedergabe von Marken, Produktnamen, Gebrauchsnamen,
Handelsnamen, Warenbezeichnungen u.s.w. in diesem Werk berechtigt auch ohne besondere
Kennzeichnung nicht zu der Annahme, dass solche Namen im Sinne der Warenzeichen- und
Markenschutzgesetzgebung als frei zu betrachten wären und daher von jedermann benutzt
werden dürften.

Coverbild: www.purestockx.com

Verlag: VDM Verlag Dr. Müller Aktiengesellschaft & Co. KG
Dudweiler Landstr. 125 a, 66123 Saarbrücken, Deutschland
Telefon +49 681 9100-698, Telefax +49 681 9100-988, Email: info@vdm-verlag.de
Zugl.: Portland, Portland State University, Diss., 2004

Herstellung in Deutschland:
Schaltungsdienst Lange o.H.G., Zehrensdorfer Str. 11, D-12277 Berlin
Books on Demand GmbH, Gutenbergring 53, D-22848 Norderstedt
Reha GmbH, Dudweiler Landstr. 99, D- 66123 Saarbrücken
ISBN: 978-3-639-09484-8

Imprint (only for USA, GB)
Bibliographic information published by the Deutsche Nationalbibliothek: The Deutsche
Nationalbibliothek lists this publication in the Deutsche Nationalbibliografie; detailed
bibliographic data are available in the Internet at http://dnb.d-nb.de.
Any brand names and product names mentioned in this book are subject to trademark, brand or
patent protection and are trademarks or registered trademarks of their respective holders. The use
of brand names, product names, common names, trade names, product descriptions etc. even
without
a particular marking in this works is in no way to be construed to mean that such names may be
regarded as unrestricted in respect of trademark and brand protection legislation and could thus
be used by anyone.

Cover image: www.purestockx.com

Publisher:
VDM Verlag Dr. Müller Aktiengesellschaft & Co. KG
Dudweiler Landstr. 125 a, 66123 Saarbrücken, Germany
Phone +49 681 9100-698, Fax +49 681 9100-988, Email: info@vdm-verlag.de

Copyright © 2008 VDM Verlag Dr. Müller Aktiengesellschaft & Co. KG and licensors
All rights reserved. Saarbrücken 2008

Produced in USA and UK by:
Lightning Source Inc., 1246 Heil Quaker Blvd., La Vergne, TN 37086, USA
Lightning Source UK Ltd., Chapter House, Pitfield, Kiln Farm, Milton Keynes, MK11 3LW, GB
BookSurge, 7290 B. Investment Drive, North Charleston, SC 29418, USA
ISBN: 978-3-639-09484-8

ABSTRACT

An abstract of the dissertation of Longjun Liu for the Doctor of Philosophy in Systems Science presented November 29, 2004.

Title: Employing simulation and optimizers to optimize experimental design and structural topology

The topology design of structures has a major impact on their cost and performance. However, currently available methods for topology optimization cannot be effectively applied to large-scale structures. This research developed two new methods for this purpose.

The first method, called hierarchical, interactive, and metamodel-based optimization (HIMO), combines a sizing optimizer with a metamodeling technique. At the lower level, for each candidate topology design, a sizing optimizer finds feasible and optimal solutions in terms of sizing variables (plate thickness in continuum structures). All performance constraints such as stress, displacement, stability, etc., are handled only at this level. At the upper level, a metamodel is built to fit all the optimal solutions found at the lower level. The metamodel is then used to find the optimal topology design. Only the objective function (e.g. weight) is approximated. The number of topology design variables is much smaller than required by other topology optimization methods, thus large-scale structural systems can be optimized. HIMO was applied to two design projects, resulting in 18% and 36% weight savings, and significant reductions in manufacturing cost.

The second method, called sizing optimizer for topology optimization (SOTO), directly uses a sizing optimizer to optimize topology as well as thickness, using finite element analysis as the simulation tool. The thinner elements are gradually deleted to achieve improved or even optimal topology. The design problem is then reformulated with much fewer design variables for final sizing optimization. For complex 3D problems, an associated partial ground structure approach is also proposed. SOTO was applied to several numerical test problems, and then to a real design project, demonstrating the applicability and efficiency of SOTO.

In order to improve metamodeling used in HIMO, through more effective sampling, 18 experimental design methods were compared and some fundamental issues in experimental design for computer experiments were explored. The results show that sample size has more impact on metamodel accuracy than the particular experimental design methods utilized, and also that more uniformly distributed sampling does not necessarily lead to more accurate metamodels. Some experimental design methods often appear to be better than the others.

To my parents, sisters, wife, and son for their support and encouragement.

ACKNOWLEDGEMENTS

The author wishes to thank Professors Bradford Crain, Faryar Etesami, Andy Fraser, Roy Koch, George Lendaris, Wayne Wakeland, Hormoz Zareh, and Martin Zwick of Portland State University for their advice and help during his study and/or this research. Especially, the chair of the dissertation committee, Professor Wakeland spent a lot of time to help re-organizing and revising the dissertation and presentation and modifying related papers published; Professor Etesami discussed some statistics areas with the author and helped modifying the dissertation and a related paper published.

Many thanks also go to Mr. Greg Saxton, PE, the Chief Engineer at Gunderson Inc., for his support and encouragement in many years.

It is appreciated that Professor Eschenauer of University of Siegen sent his papers and his book on topology optimization and Professor Papadrakakis and his colleagues at National Technical University, and many other researchers sent their papers to the author.

The author is grateful to Professors Theodore Allen and Bill Notz of Ohio State University, Drs. Vladimir Balabanov and Gary Vanderplaats of Vanderplaats Research and Development, Dr. Selden Crary of Crary Group (Design of Experiments), Professor Kai-Tai Fang of Hong Kong Baptist University, Drs. Anthony Giunta and Brian Rutherford of Sandia National Laboratory, Professor Johann Sienz of University of Wales Swansea, Professor Timothy Simpson of

Pennsylvania State University, Professor Colby Swan of University of Iowa, and Professor Bo-Ping Wang of University of Texas at Arlington who read one or more related papers and responded with similar observations, explanations, encouragement, and/or advice.

Gratitude also goes to Multnomah County Library and Portland State University Library that have obtained many research papers for the author.

TABLE OF CONTENTS

LIST OF TABLES

LIST OF FIGURES

1. Introduction

The layout or topology design of structural systems has a great impact on their performance. Better designs can result in more efficient products and great benefits for both system builders and their customers. In recent years, research in this important field has led to significant advances. However, practical approaches and techniques to address many real design needs have yet to be fully developed, especially for large-scale complex structures.

The aim of structural topology optimization is to redistribute material within the design space in an iterative and systematic manner in order to reach a structural topology that meets the design requirements with the least weight, for example.

There has been extensive research in structural topology optimization in last three decades (Ohsaki and Swan, 2002; Papalambros, 2002; Xie, et al, 2002; Rozvany, 20011,2; Papadrakakis, et al. 2001; Vanderplaats, 1999; Sobieszczanski-Sobieski and Haftka, 1997; Bendsoe and Kikuchi, 1993; etc.). Two major families of the methods have been developed. They are h-methods based on mathematical homogenization, and e-methods based on heuristic algorithms. Details of these methods are provided in Chapter 2. However, despite the availability of h-methods and e-methods, many challenges remain.

"Application of optimization into practice is comparatively modest and reserved for special problems solved by optimization experts. ... The major reason why so little of the vast research output in structural optimization has filtered through to design practice is that very little of it satisfies the specific needs of its

1

potential users." (Cohn, et al. 1994. Today, the situation is somewhat better, but many problems still remain.

The following paragraphs describe various problems associated with the optimization methodologies in general, with the h-methods in particular, and with employing metamodeling for topology optimization. Problems associated with e-methods are then discussed. The sources for this chapter include the literature* (the major references are shown at the end of this section and Reference at the end) plus this author's observations in the design and optimization of real structures.

The following list of nine topics represents the problems associated with different methodologies for structural topology optimization.

1) Practicality (associated with h-methods)

2) Local constraint (associated with employing metamodeling)

3) One shot approach

4) Practical needs and examples

5) 3D design needs

6) Structural system configuration

7) Mathematical issues

8) Initial plate thickness concerns (associated with e-methods)

9) Ground structure approach issues.

1) Practicality

Many optimization methodologies have been proposed, but many methodologies pursue technologically unrealistic goals or constraints (Rozvany and Zhou, 1996), and thus are not readily available for practicing design engineers. One of the problems associated with the basic h-methods is that before starting optimization, the user must specify the volume portion of the final structure vs. the whole volume of the reference or design domain. Further, the volume portion is the only constraint employed. The objective is to minimize compliance. In many real cases, however, the designer does not know the portion before optimization, and, the most important constraint is stress, not the volume portion. Also, the objective is often to minimize the weight, not compliance. Many published optimization examples address only academic rather than real world problems. Some researchers search for mathematically elegant approaches that may or may not address the practical techniques needed by designers.

The practical techniques addressing real design needs such as minimizing weight with stress and other local constraints, are not yet mature, especially for large-scale 3D structural systems (e.g., Schramm, et al. 1999). While scholars search for the perfect methodology, design engineers must make do with the available tools. Applying response surface optimization or metamodel optimization is one such tool.

2) Local constraint

Some researchers have used metamodel optimization for structural topology optimization. The typical approach has been to approximate the constraints and the objective functions simultaneously.

Some people developed multi-level approaches for shape or topology optimization. Wang et al. (1997) proposed a two-level decomposition method with sizing variables handled at the first level and shape variables at the second level, but all the constraints (above some critical values) are handled at both levels. This is likely to be computationally problematic for the topology optimization of real world structures.

If all the constraints, especially local constraints such as stress are involved in topology optimization, as seen in many topology optimization methods, a large number of topology design variables must be handled, making it very difficult or too expensive to optimize large-scale structural systems comprising many elements or subsystems. Thus, many methods currently in use focus on optimizing components at the 2D component level, and are not yet used for large-scale 3D structural systems. A method that requires significantly fewer design variables would be able to solve large-scale 3D problems.

3) One shot approach

Many researchers in topology optimization pursue all-at-once or one-shot approaches, i.e. sending the request only once to computers, expecting it to finish the whole task, rather than taking a more interactive approach. Many treat the optimization task as a purely mathematical problem, and do not take into account practical domain knowledge and experience available to designers. The automatic approaches are "purer", but might not be practical for many topology optimization problems. Quite often, many critical aspects associated with providing practical solutions are neglected. The automatic methods can only find optimal solutions for over-simplified problems in many cases. On the contrary, through interaction between designers and computers, metamodel-based optimization for structural topology optimization provides an effective way to combine the knowledge and intelligence of the engineer with the computational power of the computer, thereby creating better designs that neither could have done alone.

4) Practical needs and examples

"Optimization could become more attractive to practicing designers if more concrete examples of its application were available, especially for realistic structures, loading conditions and limit constraints. Broadening the practical range of structural optimization application is likely to be facilitated by addressing the needs of designers and appealing to their desire for problem-driven solutions" (Cohn, et al. 1994).

5) 3D design needs

At present, not much study is devoted to 3-D optimal topology design of structures. For these types of problems, the main difficulty when an h-method is used is that the orientation of the material voids is more complicated than in the 2-D case. The fully stressed design technique does not have this difficulty. While large-scale two-dimensional structures can be clearly be treated with both discrete and continuum topology optimization methods, the computational costs for obtaining optimal solutions for real 3D structures is still rather high. Gains made in approximation methods for sensitivity analysis and reanalysis, new approaches of mathematical programming, and parallel/distributed computing, and design and analysis of computer experiments (DACE) need to be harnessed so that three-dimensional structures can be practically analyzed.

6) Structural system configurations

Impressive advances have been made in optimal topology design of relatively simple structures. However, optimal configuration design methods for general structural or mechanical systems that may contain many parts have not been practical yet. It is a great challenge to find the optimal proportioning of components and subsystems that work together to achieve the overall system design goals.

7) Mathematical issues

Both discrete and continuum topology optimization of structures involve challenging mathematical issues. Combination of the methodologies developed in mechanical engineering, civil engineering, computer science and applied mathematics is highly recommended for further development and improvement of structural optimization methods (Ohsaki and Swan, 2002).

The above sections outlined several problems associated with the first family of the structural topology optimization, h-methods, as well as problems associated with structural optimization in general.

There are also problems associated with the second family of methods that are heuristic or interactive in nature—the e-methods. Two key problems are discussed below.

8) Initial plate thickness

Many heuristic or interactive methods have been proposed, e.g. ESO (evolutionary structural optimization, Xie, Y., et al., 2002). ESO has been receiving increased attention. The advantages and disadvantages of ESO have been discussed by many researchers (e.g. Rozvany, 20012; Zhou and Rozvany, 2001; Bulman et al., 2001; Querin, Et al., 2000; Reynolds, et al., 1999). Nevertheless, the important questions regarding the initial and optimal thickness have not been addressed. How should the initial thickness be decided? Does this choice affect the optimality of resulting topology? Could this initial selection limit the ability to locate the optimal topology and optimal thickness? If so, how should one specify an initial thickness to assure that the optimal values are found?

9) Ground structure approach issues

Ground structure approach is widely used in structural topology optimization. The design space is filled with material or finite elements, and then the optimization approach decides whether a portion of the space should keep or eliminate material or the elements. For large real structures, this effort can be very time consuming.

To address the problems mentioned above, this research proposes two new methods for topology optimization of real structures. The first method combines a sizing optimizer and metamodeling. The second method directly uses a sizing optimizer to optimize topology as well as sizing (plate thickness in continuum structures). In order to make the first method more effective and efficient, some issues regarding experimental design for metamodeling were investigated. The details on the problem background and the investigation are presented in Appendix. Eighteen design types were compared for prediction accuracy. Two important findings from the investigation regarding sample size effect and sampling uniformity were made.

The dissertation is organized as follows. Chapter 2 describes the development in structural topology optimization, metamodeling or computer experiments, and some global optimization methodologies. Chapter 3 provides research goals and methods including the procedures employed by the new methods HIMO and SOTO. Chapter 4 presents the experimental results of HIMO including the numerical tests and real applications. Chapter 5 presents similar results for SOTO. Chapter 6 shows the results for comparing eighteen experimental designs, the test results, and discussion on sample size effect and sampling uniformity. Chapter 7 completes the dissertation with a discussion of the results, limitations, implications, and conclusions. The details in the exploration of some important issues in experimental designs for computer experiments are provided in the appendix.

* The primary sources for this chapter include the following and more in References: Bendsoe and Kikuchi (1999), Cohn, et al. (1994), Eschenauer, et al (2001), Hassani and Hinton (1999), Ohsaki and Swan (2002), Papadrakakis et al (2001), Papalambros (2002), Vanderplaats (1999), Xie, et al (2002).

2. Development in structural topology optimization, metamodeling, and optimization methodologies

The development in topology optimization, metamodeling or computer experiments, and global optimization algorithms are presented below (This chapter is a review of the literature, not the contributions of the author).

2.1. Structural topology optimization*

Structural layout optimization has great impact on the performance of structures. It involves the most difficult design problems for creating a proper topology, layout, or configuration (these terms are used synonymously in this report). By topology of a structure, we mean the arrangement of material or the positioning of structural elements. The aim of structural topology optimization is to redistribute material within design space in an iterative and systematic manner to find a structural topology that is the best in terms of the design objectives.

There are two broad classes of topology optimization problems corresponding to two types of structures: discrete and continuum structures. In the first class, the structure consists of discrete truss and/or frame, each having well-defined length and cross-sectional properties. In the second class, the structure is a solid continuum of variable topology. The distinction between the two classes leads to fundamentally different optimization techniques and results. The optimization methods for both classes have undergone extensive research and development in the past decades.

Continuum structural topology optimization is a relatively recent development, having evolved from distributed parameter approaches to shape and topology optimization. This research focuses on the continuum class only.

Structural topology optimization in general can be thought of as the determination of the optimal spatial material distribution. For a given set of loads and boundary conditions, the problem is how to redistribute the material in order to optimize the objective functions, e.g., minimizing weight, compliance, displacement, or maximizing eigenvalues for buckling strength, etc. The problem can be considered as a point-wise material/no material problem that does not need to be represented by the topology parameters or functions. In solving the problem, both the reference domain and design domain that can be modified are determined first, followed by finite element discretization.

The methods for obtaining optimal topology vary from rigorous mathematically based methods to more engineering intuitive methods. There are mainly two families of approaches for continuum structural topology optimization: material/microstructure approaches include homogenization-based techniques (also called h-methods) and geometric/macrostructure approaches (also called e-methods).

2.1.1. Material/microstructure approaches

The goal of this family is to find a topology that optimizes a design objective subject to a prescribed amount of structural material. Topology is often chosen to distribute the material evenly in some porous, microstructural form over the

10

admissible design domain. The optimization process determines whether each element in the continuum should contain material or not. The density of material within each finite element is used as a design variable, defined between the limits of 0 (void or very weak material) and 1 (solid material).

The homogenization family of the methods or h-methods include the following, corresponding to different material models: a) the classic homogenization method involving square microcells with rectangular voids; b) the artificial material or SIMP methods (solid isotropic microstructure with penalty); c) the rank-1 and rank-2 methods

For these methods, the distribution of material throughout the structure is optimized using an optimality criteria procedure. The objective function is the mean compliance of the structure with an equality constraint on the volume fraction, as mentioned in Chapter 1. Finite element analysis is the simulation tool. Each element consists of a cellular material with a specific microstructure. The geometric parameters of these microstructures are the design variables of the optimization problem. The problem is solved in a fixed domain so that the finite element model used in the analysis does not need to be altered during optimization process.

Homogenization is a mathematical theory with applications in engineering problems defined on domains with regular microscopic heterogeties. In place of homogenization theory, artificial material models also can be used. These material models are simple, and usually result in more practical designs; however, the values of the objective functions become distorted. Homogenization can be

11

employed for the evaluation of results from such models and to compare different solutions.

The structural optimization by homogenization methods involves two steps: homogenization and optimization. In the homogenization part, it is established an elastic constitutive relationship as a function of the size parameters of cellular microstructures that comprise the presumed material of each element. For the material model with rectangular holes, the homogenization equation is solved for a set of values of dimensions of the cells and then each element of the homogenized elasticity matrix is expressed as an explicit polynomial in terms of the values. The elasticity matrix is a function of the geometric micro-dimensions. Artificial material model is the alternative to the model above. For rank-layered material models, analytic solutions exist for the homogenization equations.

In the optimization part, the total potential energy is the objective function to be minimized. The design variables are the geometric parameters of presumed material models in finite elements. The volume of material is the global constraint that needs to be satisfied. The optimization algorithm is the optimality criteria method.

2.1.2.   Geometric/macrostructure approaches (e-methods)

The structure consists of solid, isotropic or anisotropic material. The topology is changed by two different procedures: degenerating and/or growing a structure; inserting holes in a structure or the bubble method. The techniques that use degenerating and/or growing a structure include variable thickness sheets;

SHAPE methods; optimization by simulation of biological growth: CAO-/SKO-method; optimization by bi-directional evolutionary structural optimization method (ESO, BESO); optimization by metamorphic development method. Some people classify e-methods or the evolutionary family of methods as hard-kill methods and soft-kill methods.

This family originated from fully stressed design techniques. The basic principle is that in a design domain, material that is not structurally active — having low stress or low strain energy density — is used inefficiently and can be removed. The removal process can be simulated by either varying the elastic modulus as a function of the stresses or strain energy density, or by deleting from the structure the space occupied by the zones with low stresses/strains. Finite element analysis is again used as the simulation tool.

One of the important e-methods is called the evolutionary structural optimization (ESO) methods. Recent developments allow for material addition as well as removal. ESO is capable of solving size, shape, and topology structural optimization for static, dynamic, stability and heat transfer problems or combinations of these. ESO methods have been implemented in some FEA codes. The details of the basic ESO are described below (Xie, et al, 2002).

To make efficient structures, fully stressed design is pursued. By gradually removing material with lower stress, the stress level becomes more and more uniform throughout the new structure. The procedure follows.

At the beginning, the design domain is discretized as a finite element mesh. Loads and boundary conditions are applied. A finite element code is run and

13

stress is found. The stress of the elements is compared with the maximum stress of the whole structure s_{max}. The elements with the stresses below some rejection criterion are eliminated. After each finite element analysis, all of the elements that satisfy the following condition are removed from the model:

$$s / s_{max} < RRi \quad \text{(current rejection ratio), } i = 0,1,2,...$$

The cycle of finite element analysis and element removal is repeated using the same value of RRi until a steady state is reached, when there are no more elements being deleted at the current iteration. Then, an evolutionary rate (ER) is added to the rejection ratio:

$$RR_{i+1} = RR_i + ER \qquad\qquad i = 0,1,2,...$$

With the new rejection ratio, the cycle of finite element analysis and element removal takes place again until a new steady state is reached. The process continues until a desired optimum is reached. The topology at any steady state may be chosen as the final design. The final result may not be optimal, but the process provides many improved designs.

For the two parameters, the initial rejection ratio RR and evolutionary rate ER, the typical value of 1% has been used as the initial values for many test problems. For some problems, however, much lower values need to be used. The two parameters should be tried for any new models. If too much material has been removed from the structure within one iteration of one steady state, smaller values should be used for RR or ER.

2.1.3. Hybrid methods

Beyond the two major families of the methods mentioned above, there are some hybrid methods or H/e methods. These hybrid methods contain attributes of both h- and e-methods in different degrees. The first method of this family, developed by Fuches et al, characterizes the topology material in a manner similar to that of the original microcell model of Bendsoe and Kikuchi, using the 'Aboudi-cell' methods. The constrained adaptive topology optimization (CATO) was introduced by Hinton et al. This method uses an artificial material model and has been extended to use a variation of the classical microcell model. The aim is to update the density parameters for each element within a given design domain using a mass preserving scheme that may change during the iterative improvement.

2.2. Design and analysis of computer experiments (DACE)**

Deterministic computer simulation of physical phenomena is becoming widely used in science and engineering. A computer experiment is a set of runs of the code with different inputs for different runs. A feature of many computer experiments is that the output is deterministic, i.e. rerunning the code with the same inputs gives identical observations.

Goals in computer experiments include prediction approximation, optimization, uncertainty analysis, calibration, and integration. The variables in computer experiments are classified as control variables, environmental variables (noise variables), and model variables (model parameters).

Deterministic computer experiments differ substantially from the physical experiments performed by agricultural and biological scientists of the early 20th century. Various features of computer experiments include: 1) adequacy of the model is determined solely by systematic bias; 2) using randomization to avoid potential confounding of treatment variables due to run order is irrelevant; 3) blocking the runs into groups that represent runs on experimental units more nearly alike is irrelevant; 4) replication of runs is not meaningful.

Computer experiments involve two major areas: experimental design, model selection and building. Major statistical approaches for DACE are classified as approaches based on Bayesian statistics and frequentist approaches based on sampling techniques by some statisticians.

2.2.1. Experimental design

An experimental design represents a sequence of experiments to be performed, expressed in terms of factors or design variables set at specified levels. An experimental design is represented by a matrix, X, where the rows denote experiment runs, and the columns denote particular factor settings.

2.2.1.1. Design criteria or measures of merit for evaluating experimental designs

Selecting the appropriate design is essential for effective experimentation: the desire to gain as much information as possible about the relationships between input and output is balanced against the cost of experimentation. Several criteria and measures of merit are available and useful for evaluating and comparing experimental designs.

2.2.1.1.1. Bayesian Criteria

Koehler and Owen (1996) mentioned four design optimality criteria for use with computer experiments: entropy, mean squared-error, maximin distance, and minimax distance. Entropy designs maximize the amount of information expected for the design while mean squared-error designs minimize the expected mean squared-error. Minimax distance designs ensure that all points in the sampling area are not too far from a design point. Maximin designs pack the n design points, with their associated spheres, into the design space, with maximum radius. Parts of the sphere may be out of the space, but the design points must be in the space.

2.2.1.1.2. Discrepancy

Another type of criteria is discrepancy that is used to measure non-uniformity of designs in the space. The Lp discrepancy can be defined as

$$D_p(P_n) = \left[\int_{C^s} |F_n(x) - F(x)|^p dx \right]^{\frac{1}{p}} \tag{1}$$

F (x) is the uniform distribution function on C^s (design space). The popular L infinity discrepancy obtained by taking p = infinity is called the star discrepancy, or discrepancy for simplicity. This is probably the most commonly used measurement for discrepancy and can be re-expressed as follows:

$$D(P_n) = \sup_{x \in C^s} |F_n(x) - F(x)|$$

(2)

This discrepancy has been universally accepted in the quasi-Monte Carlo methods and number-theoretic methods. In fact, this discrepancy is the Smirnov-Kolmogorov statistic for goodness-of-fit tests. One disadvantage of the discrepancy is that it is expensive to compute.

An analytic formula for calculating the L2 discrepancy is given by:

$$(D_2(P_n))^2 = 3^{-s} - \frac{2^{1-s}}{n} \sum_{k=1}^{n} \prod_{l=1}^{s} (1 - x_{kl}^2) + \frac{1}{n^2} \sum_{k=1}^{n} \sum_{j=1}^{n} \prod_{i=1}^{s} [1 - \max(x_{ki}, x_{ji})] \quad (3)$$

$x_k = (x_{k1}... x_{ks})$; n is the number of samples; s is the number of dimensions; P_n: is a set of n points. The L2 discrepancy is much easier to calculate numerically than the L infinity discrepancy. Unfortunately, the L2 discrepancy exhibits some disadvantages, as pointed out by Hickernell (1998). To overcome these disadvantages, Hickernell proposed three new measures of uniformity that are also related to the L2 norm: the symmetric L2 discrepancy (SL2), the centered L2 discrepancy (CL2), and the modified L2 discrepancy (ML2).

The centered L2 discrepancy is:

$$(CL_2(P_n))^2 = \left(\frac{13}{12}\right)^s - \frac{2}{n}\sum_{k=1}^{n}\prod_{j=1}^{s}\left(1+\frac{1}{2}|x_{kj}-0.5|-\frac{1}{2}|x_{kj}-0.5|^2\right) + \frac{1}{n^2}\sum_{k=1}^{n}\sum_{j=1}^{n}\prod_{i=1}^{s}[1$$
$$+\frac{1}{2}|x_{ki}-0.5|+\frac{1}{2}|x_{ji}-0.5|-\frac{1}{2}|x_{ki}-x_{ji}|\Big]$$

(4)

The symmetric Lp discrepancy and its measure of variation are modified such that they are invariant if x_{kj} is replaced by 1 -x_{kj} for any j, $1 \le j \le s$. The formula for the symmetric L2 discrepancy follows.

$$(SL_2(P_n))^2 = \left(\frac{4}{3}\right)^s - \frac{2^{1-s}}{n}\sum_{k=1}^{n}\prod_{j=1}^{s}(1+2x_{kj}-2x_{kj}^2) + \frac{2^s}{n^2}\sum_{k=1}^{n}\sum_{j=1}^{n}\prod_{i=1}^{s}[1-|x_{ki}-x_{ji}|]$$

(5)

Finally, for the modified Lp discrepancy, the projection uniformity over all subdimensions can be considered. The modified L2 discrepancy also has a formula:

$$(MP_2(P_n))^2 = \left(\frac{4}{3}\right)^s - \frac{2^{1-s}}{n}\sum_{k=1}^{n}\prod_{l=1}^{s}(3-x_{kl}^2) + \frac{1}{n^2}\sum_{l=1}^{n}\sum_{j=1}^{n}\prod_{i=1}^{s}[2-\max(x_{ki},x_{ji})]$$

(6)

2.2.1.1.4.   Orthogonality, rotatability, minimum variance, and minimum bias

To facilitate efficient estimates of parameters, four desirable characteristics of an experimental design are orthogonality, rotatability, minimum variance, and minimum bias. A design is orthogonal if, for every pair of factors x_i and x_j, the sum of the cross products of the N design points:

$$\sum_{u=1}^{N} x_{tu} x_{ju}$$

(7)

is zero. For a first-order polynomial model, the estimates of all coefficients will have minimum variance if the design can be configured so that

$$\sum_{u=1}^{N} x_{tu}^2 = N$$

(8)

The variance of the prediction will also have constant variance at a fixed distance from the center of the design, and the design will also be rotatable.

2.2.1.1.5. D-optimal and D-efficiency

A design is said to be D-optimal if |X'X|/np is maximized where X is the expanded design matrix which has n rows (one for each design setting) and p columns (one column for each coefficient to be estimated plus one column for the overall mean). The D-efficiency statistic for comparing designs, as shown below, compares a design against a D-optimal design, normalized by the size of the matrix in order to compare designs of different sizes.

$$D - efficiency = (|X'X|_{design} / |X'X|_{D-optimal})^{1/p}$$

(9)

Other statistics for comparing designs such as G-efficiency, Q-efficiency, and A-optimality have also been formulated; see Myers and Montgomery, 1995.

There are many voices in the discussion of the relative merits of different experimental designs, and it is therefore unlikely that we have presented them all. The opinions on the appropriate experimental design for computer analyses vary; the only consensus reached thus far is that designs for non-random, deterministic computer experiments should be "space filling," which treats all regions of the design space equally.

2.2.1.2. Unsaturated/saturated and supersaturated designs

In many cases, the primary concern in the design of an experiment is its size. Most designs are unsaturated in that they contain at least two more design points than the number of factors. A saturated design is one in which the number of design points is equal to one more than the number of factor effects to be estimated. Saturated fractional factorial designs allow unbiased estimation of all main effects with the smallest possible variance and size (Box, Hunter et al., 1978). The most common examples are the Plackett-Burman two level design and Taguchi's orthogonal arrays. For estimating second order effects, small composite designs have been developed to reduce the number of required design points. A small composite design is saturated if the number of design points is $2k+k(k-1)/2+1$ (k: the number of coefficients to be estimated for a full quadratic model). Myers and Montgomery (1995) note that recent work has suggested that these designs may not always be good. Finally, in supersaturated designs the number of design points is less than or equal to the number of factors (Draper and Lin, 1990).

Using unsaturated designs for predictive models is most desirable, unless running the necessary experiments is prohibitively expensive. When comparing experiments based on the number of design points and the information obtained, D-optimal and D-efficiency statistics are often used.

21

2.2.1.3. Latin hypercube design

Latin hypercube sampling was introduced by McKay, Beckman and Conover in "the first paper on computer experiments." The sample points are stratified on each of input axes. The stratification in Latin hypercube sampling usually reduces the variance of estimated integrals. In Latin hypercube sampling (LHS), the jth component of the ith sampled point is

$$\mathbf{X}_{ij} = \frac{\pi_{ij} - U_{ij}}{n} \tag{18}$$

where the π_{ij} is the jth element of the ith independent uniform random permutations of the integers 1 through n (n is the number of samples), and the U_{ij} is the jth element of the ith independent U [0, 1] (uniform distribution between 0 and 1) random variables independent of the π_{ij}.

A common variant of Latin hypercube sampling has centered points:

$$\mathbf{X}_{ij} = \frac{\pi_{ij} - 0.5}{n}$$

$$\tag{19}$$

2.2.1.3. Maximin Latin hypercube design (LHMm) (Morris and Mitchell, 1995)

Morris and Mitchell extended Maximin criterion and proposed the following for optimizing Latin hypercube designs.

Criterion: Φ_p criterion (minimizing this criterion is sought within Latin hypercube design)

$$\Phi_p = \left[\sum_{i=1}^{n} J_i d_i^{-p} \right]^{\frac{1}{p}}$$

n: number of design variables

J_i: number of pairs of sites with distance of d_i

d_i: the Euclidean distance between points

p: positive parameter selected by users

Leary et al. (2003) proposed a distance related metric C as an alternative to Morris and Mitchell's. C is equal to:

$$C = \sum_{i=1}^{n} \sum_{j=i+1}^{n} (1/d_{ij}^2)$$

(22)

where n is the number of sampled points and d_{ij} is the Euclidean distance between points i and j. Minimizing this quantity is sought within Latin hypercube designs.

2.2.1.5. Uniform design (Fang, et al, 2000)

Uniform designs have been successfully applied to many fields since the 1980s. A uniform design (UD) seeks design points that are uniformly scattered on the domain. Both the Latin hypercube sampling (LHS) design and UD are space filling experimental designs. LHS is in a randomly uniform fashion and the UD is in a deterministically uniform fashion. Specifically, if the experimental domain is discrete or finite, LHS is similar to UD. When the experimental domain is continuous, LHS selects points randomly from cells, UD from the center of cells.

Furthermore, LHS requires one-dimensional balance of all levels for each factor, but UD requires one-dimensional balance and s-dimensional uniformity (s is the number of dimensions). Thus, these designs are similar in one dimension but can be very different in higher dimensions.

Fang et al. (2000) recommended several discrepancy expressions. One of them is Hickernell's centered analytical L2 discrepancy CL2. This discrepancy is one of the criteria used by the author for HIMO.

Experimental design for computer experiments is a very active research area today. Many new approaches are developed quickly.

2.2.2.   Model selection and building

Many metamodeling techniques have been studied and applied. They include response surfaces or polynomial metamodels (Montgomery and Myers, 1995); Kriging, DACE, or spatial correlation models (Sacks, et al, 1989; Simpson, et al, 2001); Radial basis function; Neural networks (NN); inductive learning (Langley and Simon, 1995); Multivariate Adaptive Regression Spline (MARS) (Friedman 1990); an accumulated approximation technique for structural optimization (Rasmussen, 1990); weighted and gradient-based approximations for use with optimization (Balling and Clark, 1992); multivariate Hermite approximations for multidisciplinary design optimization (Wang, et al, 1996); wavelet modeling (Mallet, 1998); projection pursuit regression (Friedman and Steutzle, 1981); etc.

A relative new tool for metamodeling is support vector machine that includes support vector classification and support vector regression. Some papers report that better performance was achieved for prediction with this tool than achieved by other models (e.g. Clarke, et al, 2003).

The most popular metamodel, Kriging models, are described below. It is the model used in this study. Kriging (or Bayesian) models are very flexible and powerful approach for modeling the response of computer experiments. This approach includes a linear model plus the bias of the response from the model as a stochastic process. This provides a statistical basis for estimating prediction uncertainty. There are several correlation families for representing the process. This approach interpolates observed responses as exact predictions and predicts with increasing errors as the untried point moves away from all the sampling points.

Sacks, et al (1989) suggest modeling responses of computer experiments as a combination of a polynomial function plus a bias:

$$y(x) = f(x) + Z(x) \tag{28}$$

where $y(x)$ is the unknown function for prediction, $f(x)$ is a known polynomial function, and $Z(x)$ can be the realization of a normally distributed Gaussian random process with mean zero, variance σ^2, and non-zero covariance, as one of the stochastic processes. The covariance matrix of $Z(x)$ is given by:

$$Cov[Z(x^i), Z(x^j)] = \sigma^2 \mathbf{R}([R(x^i, x^j)]$$

$$\tag{29}$$

where $\mathbf{R}$ is the correlation matrix, and $R(x^i, x^j)$ is the correlation function between any two of the sampled data points x^i and x^j. R is a symmetric matrix with 1's along the diagonal. $R(x^i, x^j)$ is selected by the user from several correlation functions that may be used. A popular Gaussian correlation function has the form:

$$R(x^i, x^j) = \exp[-\sum_{k=1}^{n_s} \theta_k |x_k^i - x_k^j|^2]$$

(30)

where θ_k are the unknown correlation parameters and the x_k^i and x_k^j are the kth components of sample points x^i and x^j.

Prediction $\hat{y}$ (x) of the response y(x) at a untried point x is given by:

$$\hat{\mathbf{y}} = \hat{\beta} + \mathbf{r}^T(\mathbf{x})\mathbf{R}^{-1}(\mathbf{y} - \mathbf{f}\hat{\beta})$$

(31)

where **y** is the column vector of the values of the response at sample points; **f** is a column vector of 1's when f(x) is taken as a constant; $\mathbf{r}^T(x)$ is the correlation vector between x and the sample points $\{x_1, x_2, ..., x_{ns}\}$.

The formulas for the items follow.

$$\mathbf{r}^T(\mathbf{x}) = [R(\mathbf{x}, \mathbf{x}^1), R(\mathbf{x}, \mathbf{x}^2), ..., R(\mathbf{x}, \mathbf{x}^{ns})]^T$$

(32)

$$\hat{\beta} = (\mathbf{f}^T\mathbf{R}^{-1}\mathbf{f})^{-1}\mathbf{f}^T\mathbf{R}^{-1}\mathbf{y}.$$

(33)

$$\hat{\sigma}^2 = \frac{(\mathbf{y} - \mathbf{f}\hat{\beta})^T \mathbf{R}^{-1}(\mathbf{y} - \mathbf{f}\hat{\beta})}{n_s}$$

(34)

The last item is the estimate of the variance from the underlying global model, where f(x) is assumed to be the constant; n is the number of samples; s is the number of dimensions. To find the θ_k used to fit the model, the following maximum likelihood estimates are maximized

$$-\frac{n_s \ln(\sigma^2) + \ln|\mathbf{R}|}{2}$$

(35)

for $\theta_k > 0$ where both $\hat{\sigma}^2$ and $|\mathbf{R}|$ are functions of θ_k. Depending on the choice of correlation function, Kriging can either provide an exact interpolation of the data, or can smooth the data, providing an inexact interpolation.

2.3. Optimization algorithms for global optimization***

The mathematical optimization problem in its general form is difficult to be solved when the objective functions are nonlinear and there are many local optima. During the last three decades there has been a growing interest in Evolutionary Algorithms (EA), the global optimization approaches based on analogies to natural processes. EA include genetic algorithms (GA), evolutionary programming (EP), and evolutionary strategies (ES).

Evolution-based systems maintain a population of potential solutions. Some selection process based on fitness of individuals and some recombination operators are used. Both GA and ES imitate biological evolution and combine the concept of artificial survival of the fittest with evolutionary operators to form a robust search mechanism. GA usually does not require evaluating derivatives, but need a large amount of function evaluations. They require no special mathematical knowledge and can be easily programmed. But, good implementations require some skill and experience.

In the following, EA methodologies will be presented, focusing on GA. Then, brief descriptions will be given to other newly developed global optimization methods.

2.3.1. Genetic Algorithms (GA)

Genetic algorithms are the best-known evolutionary algorithms. In the basic genetic algorithms, each member of a population is a binary or a real-valued string, referred to as a genotype or a chromosome. The phenotype is the value of the design variable or chromosome in real value representation. Different versions of GA have appeared in the literature in the last decade, dealing with methods for handling the constraints or techniques to reduce the size of the population.

Genetic algorithms search through the space of potential solutions to the problem. GA performs independent sampling on a large population of design solutions, then selects members of the population (highly fit designs) for survival and creates new designs by crossover (combining building blocks for different individuals) and mutation. Crossover ensures the inheritance and dominance of valuable features, whereas mutation introduces variability. Together with the selection mechanism, crossover and mutation drive the artificial evolution process towards generating continuously better solutions.

Genetic algorithms are being applied to many areas of engineering including mechanical engineering, electrical engineering, aerospace engineering, architecture and civil engineering, etc. The major aspects of the basic GA are described below, followed by introducing advanced GA.

2.3.1.1. Representation

There are many ways for the representation of the populations. Some representations can successfully lead to good solutions, while others fail to converge or take too much time to complete search. A common representation used in simple genetic algorithms is the fixed length bit string. In the case of more complex problems, a more sophisticated representation might lead to better results. Real value representation offers several advantages over binary encoding. It can lead to quicker solutions if the genetic operators are properly designed. The efficiency is increased and less memory is required. There is also greater freedom to use different genetic operators (Chipperfield, A., et al. 1995).

In solving complex design problems, a two or three-dimensional representation is often much more successful than one-dimensional representation. In other cases, variable length representations may be chosen when the genetic algorithm finds the appropriate length of the chromosome and its content at the same time.

In general, incorporating knowledge specific to the problem domain into the representation helps to guide the evolutionary process towards good solutions.

2.3.1.2. Fitness assignment and evaluation of fitness function

GA evaluates both the objective function and the fitness. The evaluation of a string (or a member in the population) refers to the evaluation of the objective function value of that string, and it is independent of the evaluation of any other string. The fitness of that string, however, is always defined with respect to other members of the current population. The fitness is used to determine the selection probability of this chromosome to participate in the generation of new chromosomes. Fitness can be assigned based on a string's rank in the population or by sampling methods, such as tournament selection.

2.3.1.3. Selection

Individuals are selected to have offspring. Those individuals with better fitness values are picked more frequently than individuals with worse fitness values. There are several ways for selection. In fitness-proportional selection, each individual has a share directly proportional to its fitness. In ranked selection, individuals are ordered according to their fitness. They are then selected using a probability based on some linear function of their rank. In tournament section, a set of n individuals is chosen from the population at random; then the best of the pool is selected.

2.3.1.4. Crossover

Crossover forms a new chromosome by combining parts of two parent chromosomes. The simplest form is called single-point crossover, in which an

30

arbitrary point on the parents is picked. The first offspring is generated by copying all of the information from the beginning of the parent A to the crossover point, and copying all the information from the crossover point to the end of parent B. The second offspring is generated by the opposite operation.

Beyond one- or multiple- point crossover, there exist more sophisticated crossover types. In a representation consisting of numerical values, the arithmetic crossover generates offspring as a component-wise linear combination of the parents. Sometimes, domain knowledge can be incorporated into producing offspring.

2.3.1.5. Mutation

Mutation is a reproduction operator that forms offspring by alternating the values of a single parent chromosome, changing one or more values in the representation or adding/deleting parts of the representation.

By using different crossover and mutation probabilities or rates for applying the genetic operators, the speed of convergence can be controlled. Crossover and mutation must be carefully designed, which affect the performance of the genetic algorithm.

2.3.1.6. Advanced genetic algorithms

Genetic algorithms work well for many practical problems. However, when it is applied to complex design problems, simple GA may converge slowly; evaluations may be computationally intensive; or GA may converge to an unacceptable local optimum. Considerable efforts have been made to improve the efficiency of GA which has resulted in advanced genetic algorithms such as parallel and distributed GA. Parallel GA utilize multiprocessors to simultaneously evaluate fitness functions so that GA can be much faster. Multi-objective GA has also been developed for multi-objective optimization. Some advanced GA feature diversity maintenance and creative design.

2.3.2.   Other evolutionary algorithms

Among all the evolutionary algorithms, genetic algorithms are the most popular methods and have many more applications. The other evolutionary algorithms include evolutionary programming, evolutionary strategies, genetic programming, etc.

Evolutionary programming works directly on the variables of the problem and uses only inheritance from one parent. It creates new individuals only by mutation. Intelligent behavior was viewed as the ability to predict the environment and to give proper responses so as to reach a certain goal.

Evolution strategies represent individuals as real-valued vectors and also create new individuals by mutation only. In contrast to evolutionary programming, evolution strategies select the parents for a new generation deterministically.

Genetic programming (GP) is a relatively new development. GP operates on computer programs represented in various forms such as trees, sequential structures, or graphs, by using modified genetic operators. GP is called "universal approximator," but it is still in its infancy for metamodeling. It might become one of the major players as approximation models after development.

2.3.3. Simulated annealing (SA)

Simulated annealing is a stochastic search method. A new solution is obtained by perturbing the current solution. If the objective function of the new solution is better than the old, it is accepted. If it is worse, it is still possible to be accepted, based on the probability being reflected in the temperature of the system. Higher temperature results in higher probability to accept worse solutions. As search goes on, the temperature is gradually lowered, so is the probability. Being able to leave from the local optima, SA has a better chance to find the global optimum. The solutions do not depend on the starting point. However, SA requires much higher computational effort than gradient methods. SA performs well on combinatorial problems.

2.3.4. Tabu search

Tabu search is a quite new adaptive strategy that was primarily designed for combinatorial problems. It can continue exploration with a hill-descending search algorithm, even without improvements in the objective function. It can too leave from local optima that may have been found but later rejected. It uses short-term memory of recent solutions and strategies to impose move restrictions.

Many variations of this method are still being explored. Like SA and GA, Tabu search provides an approach to overcoming the problem of relative optima.

2.3.5.   Hybrid optimization algorithms

Several hybrid optimization algorithms, which combine evolutionary computation techniques with deterministic procedures for numerical optimization problems have been successfully investigated. The hybrid methods include the following combinations: evolution strategies with SQP method, evolutionary programming with the direction set method (a non-gradient method), etc.

This chapter reviewed the literature in three areas relevant to this research: structural topology optimization, computer experiments, and global optimization. Of particular interest is the unsolved problems revealed by the literature. For instance, the h-method requires the user to specify the volume portion as the sole constraints, and the objective is minimizing compliance instead of weight that is needed for many real design optimization tasks. Another example follows: in ESO, there is no method provided for selecting the initial thickness, a critical parameter that may significantly impacts the final result of topology optimization. It seems that the researchers have not yet addressed these important aspects for ESO.   The research design presented in Chapter 3 addresses these and other questions raised in Chapter 1.

* The primary sources for this section (2.1.) include the following and more in References: Bendsoe and Kikuchi (1999), Eschenauer, et al (2001), Hassani and Hinton (1999), Ohsaki and Swan (2002), Papadrakakis et al (2001), Papalambros (2002), Vanderplaats (1999), Xie, et al (2002).

**The major sources for this section (2.2.) include the following and more in References: Barton (1998), Fang, et al (2000), Koehler and Owen (1996), Myers and Montgomery (1995), Sacks, et al (1989), Simpson, et al (2001).

***The primary sources for Section (2.3.) include the following and more in References: Back, et al, 1991, Chipperfield, A., et al. 1995, Papadrakakis et al, 2001, Renner and Ekart, 2003.

3.   Research plans/methods and two new approaches for structural topology optimization

In order to address the methodological problems described in Chapters 1 and 2, two new methods for structural topology optimization were developed. The first (HIMO) combines metamodeling with a sizing optimizer. The second (SOTO) directly uses a sizing optimizer to optimize topology.

During the development of these two new methods a variety of numerical tests were performed, using numerical test problems and simplified real world design problems. Based on the results of these tests, more effective and efficient procedures and implementation approaches were developed. These approaches were further tested, compared, and improved. Both HIMO and SOTO were then employed for real design optimization. Substantial savings and much better layout designs were found, showing the effectiveness and efficiency of the new methods. Also, HIMO and SOTO were compared with other topology optimization methodologies, including evolutionary structural optimization (ESO) and homogenization methods, by comparing the optimal topology designs generated by these methods with those generated by HIMO and SOTO. Several "standard" benchmark problems were used for this comparison.

Sections 3.1 and 3.2 describe in detail the two new methodologies for structural topology optimization. Test results for each method are provided in Chapters 4 and 5. Section 3.3 presents the research methods for comparing the experimental designs for metamodeling.

3.1. Combining sizing optimizer and metamodel optimization for structural topology optimization (HIMO)

HIMO is a systematic, practical, and effective approach to layout optimization of structural systems. It combines sizing optimizer and metamodel optimization through a sequence of computer experiments. It includes two levels or two stages of work.

At the lower level, corresponding to every sampled point in the topology design space a feasible and optimal thickness (in plate cases) is found among many thickness options that satisfy the stress and other constraints.

At higher level, an approximation curve or surface is built to fit all of the feasible and optimal points found at lower level. An optimizer searches the best point on this curve, which is the goal— the best topology design corresponding to the minimum weight, for example.

It is not a simple approximation at all. The following explains why, by using a very simple example — a plate structure with only one topology design variable (e.g. the number of channels) and one sizing variable that is the thickness of the plate. The objective is to minimize the weight. Corresponding to one value of the topology variable, there can be many options for the thickness. The structure with one pair of topology variable value and thickness value may form a feasible design that can meet the stress and other constraint requirements. Another pair of the values may result in an infeasible design that does not meet at least one constraint requirement.

A simple approximation is shown in Figure 1. Each data point represents one selection of the thickness corresponding to each value of the topology variable. The approximation curve is built to fit the three points and the best is searched on this simple approximation curve. Unlike this simple approximation, in Figure 2, HIMO goes through many thickness options to find feasible and optimal thickness for every option of the topology design, at the lower level and before building the metamodel. In both figures, X is the topology variable, supposing only one; Y is the objective function, e.g. weight.

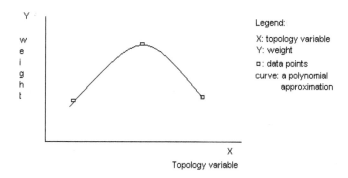

Figure 1. Simple polynomial approximation (general use of approximation or interpolation)

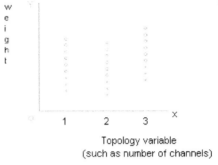

Figure 2. HIMO: lower level of HIMO

In Figure 2, for every topology design, corresponding to each column, among all the thickness options, some designs are feasible (o's), and the others are not (x's). Among all the feasible designs, there is one thickness that has the minimum objective function value. That is the optimal thickness for this value of the topology variable.

At the higher level that is for topology optimization, an optimal curve or surface is built to fit the optimal point for every value of the topology variable (see Figure 3). The topology optimization finds the best value for the topology variable from this optimal surface. This value yields the lowest objective function value and represents the best overall topology design (seems trivial in this one variable example, but is decidedly not so in real world cases)

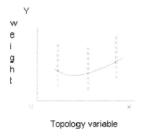

Y

weight

Topology variable

Legend:
o: a feasible design, corresponding to one thickness
x: an infeasible design, corresponding to another thickness
column: all designs correponding to one value of topology variable
curve: approximation of the optimal weight surface with every point as the optimal thickness corresponding the lowest weight of all the feasible designs. The lowest point on this curve corresponds to the optimal value of the topology variable, i.e. the optimal topology desing.

Figure 3. HIMO: both levels (The curve is the approximation of the sizing-optimal solutions. Topology optimization finds the lowest point on this curve.)

Comparing Figure 1 and 3, we can see simple approximation without searching for the optimal may find the solution opposite to or different from the true optimal design.

The procedure for HIMO is outlined below.

1) Statistical techniques for computerized experimental design are used to generate the sampling points in the topology design space, representing different configurations or layout options of the structural systems. Latin hypercube sampling, Latin hypercube plus maximizing the sum of the distances between points and/or minimizing the centered discrepancy were used in this study.

2) Finite element analyses are run for every sampled topology design. Adjustment (e.g. changing the plate thickness) is made to find feasible designs in terms of the major performance constraints such as stress, displacement, stability, as well as bound limits for sizing or property variables. At the same time, sizing-

39

optimal or thickness-optimal (in continuum cases) designs are obtained via sizing optimization. This process is repeated for every sampled point in the topology design space.

3) A metamodel representing the objective function, e.g., the structural weight, versus the topology design variables, is then fit to the results of the computer experiments. Every point on this response surface (metamodel) represents a design that is feasible and optimal in terms of the major performance constraints and lower level optimization goal (sizing optimization). Kriging models were used for metamodeling in this study.

4) The higher-level optimization is conducted to find the best points on the response surface, i.e. the optimal layout design, satisfying the bound constraints and topology constraints of the topology variables. A hybrid optimization strategy that uses genetic algorithms followed by sequential quadratic programming (SQP) was used in this study.

5) Finite element analysis with the sizing optimizer is run again to work on the topologically optimal solution just obtained to find the final result that is feasible and optimal in terms of both sizing and topology.

When the topology design space is large, sampling may be performed in multiple steps, referred to as sequential sampling. The interaction between the designer and the computer forms an interactive approach.

To successfully employ HIMO, the metamodeling strategies addressing sampling, modeling, and optimization are critical. Sampling or experimental design is the most important, since effective and efficient sampling can provide uniform and representative coverage of the space investigated. Sometimes, even without the further work of modeling and optimization, the sampling tests can direct the designers towards much better designs that might be sufficient for the task at hand. Latin hypercube designs are widely used for computer experiments. More discussion on experimental design is presented in the appendix.

Relatively new approximation models, such as Kriging models, are very powerful and extremely flexible, resulting in a much more accurate global approximation than can be accomplished using low order polynomials as has been done in the past. Details regarding Kriging models were provided in Chapter 2.

Metamodels with high nonlinearity, sometimes called "bumpy" models make the global optimization task difficult for standard mathematical programming approaches. Hybrid approaches, such as genetic algorithms followed by SQP, are often very effective for addressing difficult global optimization problems. With its combination of effective methods for sampling, modeling, and optimization, HIMO provides a systematic way for exploring the design space that is both efficient and practical.

3.2. Using a sizing-optimizer to optimize topology in conjunction with partial ground structure approach (SOTO)

The proposed heuristic approach, SOTO, directly uses a sizing optimizer to optimize topology. SOTO is different from other topology optimization methodologies, as described below.

The problem formulation is based on distributed material: plate thickness for continuum structures and cross sectional area for discrete structures, rather than density as is used by many topology optimization methods. At the beginning, finite element meshing is done to discretize the continuum structure. For simple or small problems, each element thickness can be assigned a unique property number as a design variable, and the elements are created to fill the whole reference domain or design space, following the popular ground structure approach. For more complex or large structures, variable linking and partial ground structure approach may be needed. With the latter, only part of the domain or space is intentionally filled with material or finite elements for evaluating different design options.

To optimize topology by SOTO in the continuum case, material distribution is adjusted by changing the thickness. Rossow and Taylor suggested a variable thickness sheet model for shape and topology optimization (1973). The design domain is divided into many smaller areas whose thickness are used as design variables, with a view to minimize the structural compliance for given total volume (Eschenauer, H., Olhoff, N. 2001).

The idea of structural topology optimization via distributed parameter optimization was proposed by Cheng and Olhoff (1981). They recognized that, when optimizing the spatial material distribution of the structure, regions of "zero thickness" were holes in the plate structure. Thus, by using a spatial distribution of design variables, sizing optimizers can essentially change or optimize topology as well, where topology refers to the material distribution in a structure (Ohsaki and Swan, 2002).

Generally, for continuum sizing optimization, the sizing-optimizer will make some elements thicker and the others thinner, but usually not zero--especially when there are stress constraints. When the load is large in terms of the available structure or design space, the whole plate or structure may be assigned the same thickness by the optimizer. The optimization cannot continue. In some of these cases, different initial thickness values could be used to force the optimizer to assign different thickness values to different areas so that topology optimization could proceed.

It is ideal to restrict the sizing optimizer so that it allows the thickness to only take on values between specified lower and upper bounds of the design variables; and otherwise to be zero. There is a penalty associated with the thickness outside the acceptable range. Stress constraints are relaxed if a particular design variable is assigned a value of zero. Such a methodology would be ideal for simultaneously optimizing topology and size. But such a tool does not yet exist. In the meantime, the following iterative procedure for using currently available sizing optimizers for continuum cases may be efficacious.

The principal idea is that, by gradually deleting the elements with lower thickness, the structure can be improved in terms of weight, while still satisfying the constraints. There are two phases in SOTO: topology optimization and sizing optimization.

For the problem formulation in the first phase, if a ground structure approach is adopted and finite element analysis is used, elements will fill the design space and each element is assigned a unique property number that is treated as a design variable. The initial thickness values are not necessarily different from each other. The objective function is then specified, which may be based on weight or other variables. The design variables are specified together with the lower and upper bounds. These bounds are control parameters that can have considerable impact on the optimization result. The lower bounds can be practically zero. The upper bounds can be a little higher than the desired upper bounds. Constraints, stresses, displacements, etc. are then specified.

After several (or perhaps many) cycles, the first iteration of the sizing optimizer converges and stops. The optimal thickness values are grouped into higher and lower values, using a small number as the cutoff value. The elements with the lower thickness are shown on the computer screen. If the group would break the major load path from the load to the support, then the cutoff value is too high. In this case, a lower cutoff value is tried, and the possible disintegration is checked again. If no disintegration results, the elements within the lower group are either deleted or assigned a practically zero value for possible later recovery if needed, and the stress constraint associated with these elements, if any, are relaxed. The removal of elements often significantly reduces the problem size for the following analysis and optimization, but setting the size to practically zero does not. The tradeoff is that elements with almost zero thickness could be recovered later if needed, whereas deleted elements cannot. The cutoff value decided in this fashion is in fact an upper limit. The cutoff value often needs to be smaller, especially when the recovery strategy is not used. This is explained further in Chapter 5, when SOTO is compared with ESO.

The practically zero value is very small, but not zero, in order to avoid the singularity problems of the stiffness matrix. When the maximum pivot ratio in the stiffness matrix is greater than a limit, e.g. 10^7 or smaller, depending on the program, FEA will not run, particularly for 2D problems. When the stiffness ratio of the neighboring elements is small enough, this problem is avoided. It was noticed that the stiffness ratio or the thickness difference between the neighboring elements had to be lower when the "recovery" strategy was used, i.e. not physically deleting the elements during evolution.

Next, the optimization module is re-specified to reflect the changes made to the model, e.g. the design variable changes, constraint changes, control bound changes, etc, and a second iteration is run. After the optimizer stops, the thickness cutoff value is re-evaluated and additional elements are removed. The process is repeated until an appropriate stopping criterion is met, as will be discussed shortly.

The second phase of the process now commences. Based on the topology optimization result, which is only a guideline for distributing material, the design problem is reformulated with a much smaller number of design variables for sizing optimization for final design. For instance, for a simple component, only one plate with one thickness may be selected and the topology will be based on the first phase result. The design can be made to be more manufacturable, symmetric, free of chains, and to have smoother transitions, etc. Then, the design goes through final sizing optimization with only one design variable. Ideally, the final design is optimal in terms of both topology and sizing.

Because there will be a second phase in the process, it seems that the result of the first phase, since it serves only as the guideline for the subsequent distribution of material, could contain hinges/chains or even local checkerboards, as long as no mechanism is formed to prevent FEA from running or no difficulty is caused that adversely impacts convergence. The chain or small area checkerboard problems can be resolved in the second phase. If needed, local re-meshing can eliminate the chains in the first and/or second phases. More study is needed regarding chains and checkerboards for SOTO. For symmetric problems, if the symmetric design is not achieved in the first phase, it can be enforced in the second phase.

When should the topology optimization be stopped? This is a difficult question to answer. Two criteria could be used: the thickness falls into the desired bounds and/or the major load path is not broken while trying to eliminate elements. The final design after the final sizing optimization must be within the desired thickness bounds. Ideally, it could be seen during the topology optimization phase if the final thickness would fall within the bounds. But usually the final optimal thickness is not known until the final sizing optimization is done. More study is needed to find the relationship between the thickness values at the end of the topology optimization phase and the final optimal value. It is likely that the final value is between the largest and the smallest values in the first phase. In

46

most cases, stopping early may result in a large area but with thinner plates for the final design, and opposite for stopping late. To achieve the lowest volume may depend critically upon the stopping rule. Further study is warranted.

Unlike ESO in which the initial thickness is unchanged, in SOTO, the sizing optimizer automatically redistributes the material. And, the material to be deleted is not based on the stress, but on the direct guidance of the sizing optimizer.

Quite often, in the design optimization of real structures, the desired thickness bounds are specified or restricted by other factors, and the optimal design is sought that can have the minimum weight and satisfy the constraints. Unlike h-methods, where the volume portion must be specified in advance, in SOTO the volume portion is either not specified or does not matter. The allowable thickness is usually not one value, as is specified in some ESO examples, but is actually a range, as is used or enforced in SOTO.

Like using ESO, it is necessary in using SOTO that at each step, only a small portion of the model is deleted, i.e. the deletion process should be done gradually. Although each element below the cutoff value has lower thickness values than those above the cutoff value, the lower thickness group as whole might still play a significant role. This concern may be more important when the element recovery strategy is not used. But research results to date indicate that when using SOTO, a larger number of elements could be deleted per iteration, especially at the beginning, comparing with using ESO.

Like using ESO, many additional issues might be relevant and therefore need to be addressed for using SOTO, such as mesh dependency, re-meshing or adaptation after element deletion to get smooth new boundary to avoid stress concentration, etc. In application of SOTO, a very large number of design variables together with a large number of constraints could make optimization difficult. Also, global vs. local optimization needs to be considered. It was observed, for example, that the sizing-optimizer sometimes got trapped in local optimal values. Hybrid optimization strategies, such as genetic algorithms followed by mathematical programming algorithms might be helpful. Optimality criteria methods may be helpful when there are a large number of design variables. Like ESO, SOTO can be used to optimize shape as well as topology and sizing.

3.3. Research methods for comparing the approaches for optimizing Latin hypercube designs

The research methods for comparing different design types for metamodeling are briefly described below; see Appendix for the details. The main study tools employed are simulation tests and statistical analysis based on one-way and two-way ANOVA (analysis of variance). An experimental design method is employed for sampling. Kriging models are used to approximate twenty test functions. Sampling for verification is conducted by Latin hypercube design to sample from 1000 to 10000 points within the design space. The accurate responses and those generated by Kriging models are compared to find root mean square errors (RMSE), maximum errors, and relative RMSE (RMSE divided by mean of the responses of all the response for verification). This process is repeated at least ten

48

times for the same pair of dimension and sample size. The above process is repeated for each design type. Next, the sample size is increased and the process is repeated. Finally, the entire process is repeated for another function. The numbers of dimensions (n) tested are 2, 5, and 10. The sample size varied from 4 to 100.

Chapter 4 provides detailed test results of HIMO. Chapter 5 provides detailed test results for SOTO. Chapter 6 provides detailed test results for comparing the experimental designs.

4. Test results for HIMO--Combining sizing optimizer and metamodeling for structural topology optimization

4.1. Pilot Numerical Tests with HIMO

As an initial trial to see the applicability and potential of HIMO, pilot tests were done to a part of a real structure. This simple assembly is the bottom section of a vehicle end assembly, consisting of an end sheet reinforced by two channels and a corner post. For finite element analysis (FEA), only half is modeled because of the symmetry of the structure, loading, and constraints. The FEA model (for plot only) is shown below. The border is pinned so that all the translation movement is restricted, except the symmetric center where the symmetric constraint is applied to simulate the other half. To be able to see the response surface, only two variables are selected: the depths of the two channels (along X-direction). The objective function is the weight of the assembly. The layout question is what combination of the two channel depths results in the lowest weight of the feasible structure that meets the stress requirement.

The half model dimensions (in inches) are the following: end sheet height, 22.25; width, 20; corner post width, 2; channel height, 5.5; spacing between channels and top/bottom spans, 3.75. The stress limits are -50 ksi, 50 ksi. The thickness bounds are 0.1046 and 0.5 inches. Young's modulus is 29000 ksi, and Poisson ratio is 0.3.

a)

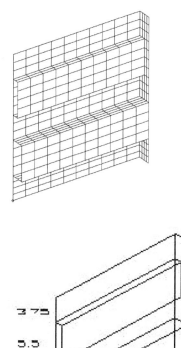

b)

Figure 4. FEA model used for pilot tests (a: model; b: dimensions)

After the problem was defined, an initial model was created based on the topology values similar to the real design. The feasible solution was found to meet the stress requirement by manual adjustment, and sizing-optimal solution was found by sizing optimization module in MSC/NASTRAN for Windows (the finite element analysis or FEA program). Then, an experimental design by Latin hypercube sampling was conducted to sample 6 points from the design space set by the bounds (See Table 1). Six points were selected based on the requirements of quadratic polynomial fitting. Six finite element models were then created and optimized for the minimum weight by changing the sizing variables or by changing the thickness of the end sheet plate, the channels and the post, subject to the stress constraints.

This end structure has several unique features. Usually, when thicker plates are selected, the stress will go down. That does not necessarily happen to this structure. If one adds thickness to one part, e.g. a channel, the stress in the end sheet could go up, instead of down as one would think. Manual adjustment may make things worse and worse (higher and higher weight). It was not until the sizing optimizer was employed that this phenomenon became apparent. The sizing optimizer finds the feasible and optimal result at the same time, much faster than manual adjustment. The six results correspond to the optimal weight of the six topology designs, each of which is optimal in terms of sizing.

A Kriging model was fit to the data to obtain the approximate weight response against the topology variables. An SQP optimizer was employed to find an optimal solution (design). Based on this new topology, another FEA model was created and sizing-optimized. Its weight was lower than any design sampled, reduced by 33% compared to the initial model with sizing optimization, and reduced by 41% compared to the design without sizing optimization. The larger portion of the savings is due to topology optimization.

Next, 15 more samples were evaluated with the same procedure: using FEA to find sizing-optimal results, fitting the Kriging model to the data, and optimizing by SQP for the optimal topology. The optimal result is better than any design already tested for building the response surface. The solution went through FEA sizing-optimization and the best result in terms of both sizing and topology was found, which at this point was reduced by 45% compared to the initial model with sizing optimization, and reduced by 51% compared to the model without sizing optimization. Again, topology optimization plays the more important role.

Figures 4-6 show the response surfaces for two topology variables, representing the models from 6 samples, 15 samples, and all 25 samples including all of the tests and the optimal solutions. The surfaces in Figure 6 and 7 are so bumpy that it may be hard for quadratic polynomial models to approximate them. The model based on 6 samples (Figure 5) was cross validated by itself (the first group); about 10% were left out for building the models. The model based on 15 samples (Figure 6) was cross validated by itself (the second group) and validated by the first group (6 samples). The empirical root mean squared error (RMSE), maximum error, and relative RMSE (RMSE divided by the mean weight) with zero and first order regression polynomials and Gaussian correlation models in the Kriging models are provided in Table 2. Subsequently, the six sample group with the lowest C (see Chapter 2, p39) among 5000 LHS sample groups was found, and the FEA models corresponding to the six tests were optimized in terms of sizing. A Kriging model was built from the data, and then was optimized by SQP. The resulting surface plot is shown in Figure 8.

Table 1. Topology designs and weights

Sampling points	Order	X1	X2	weight (lb)	Savings (%) based on 48.06
Initial model before size optimization	1	1.0	1.0	48.060	
Initial model after size optimization	2	1.0	1.0	41.590	
Group 1					
6 points	1	0.5	1.5	41.355	
	2	0.0	2.0	37.040	
	3	1.75	0.75	37.851	
	4	1.25	0.5	40.432	
	5	0.75	0.25	53.086	
	6	1.5	1.0	37.178	
	optimal	1.08	2.0	28.202	41
Group 2					
15 points	1	1.25	0.25	43.119	
	2	0.75	2.0	45.171	
	3	1.0	1.5	31.889	
	4	2.0	1.25	49.470	
	5	1.75	1.0	35.869	
	6	0.5	1.5	41.355	
	7	1.5	0.75	38.854	
	8	1.0	0.25	54.538	
	9	1.5	0.5	38.773	
	10	0.5	0.5	56.222	
	11	0.0	1.75	40.735	
	12	1.5	1.75	25.423	
	13	0.25	0.0	54.514	
	14	1.25	0.75	40.239	
	15	0.25	1.25	47.469	
	optimal	2.0	2.0	23.527	51
Group 3					
6 points	1	1.0	2.0	32.468	
	2	0.5	1.25	48.216	
	3	0.75	0.0	51.321	
	4	0.25	0.5	54.917	
	5	1.5	1.5	30.203	
	6	1.75	1.0	37.359	
	optimal	2.0	2.0	23.527	51

Table 2. Validation result

	RMSE	Max error	Rel-RMSE
Group 1			
Cross-validate using group 1			
regpoly1	4.95	9.43	12.0
regpoly0	6.77	13.9	16.4
Group 2			
cross-validate using group 2			
regpoly2	9.6	20.0	23.0
regpoly1	6.07	15.3	14.6
regpoly0	5.61	10.5	13.5
validate using group 1			
(model based on group 2 only)			
regpoly2	3.2	7.08	7.8
regpoly1	2.97	6.56	7.2
regpoly0	3.76	8.3	9.1
Group 3 (new group)			
cross-validate using group 3 (new group)			
regpoly1	4.8	7.9	11.2
regpoly0	4.11	5.7	9.7
validate using group 1			
regpoly1	3.6	7.5	8.75
regpoly0	6.45	14.1	15.7

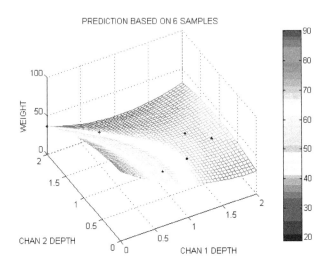

Figure 5. Response based on 6 samples

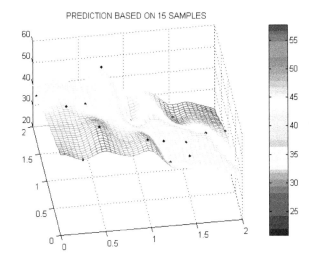

Figure 6. Response based on 15 samples

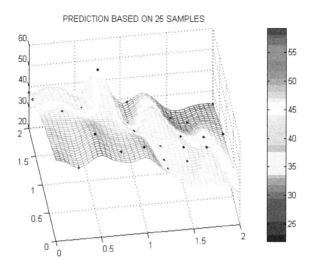

Figure 7. Response based on 25 samples

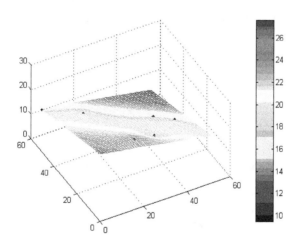

Figure 8. Response based on 6 new samples

The cross-validation by this group and validation using group 1 data mentioned above show the lower relative root mean square error (RMSE) values (9.7 % vs. 12%). RMSE, maximum error, and relative RMSE for cross-validation are listed in Table 2. The optimizer found the global optimum design (2.0, 2.0) with the minimum weight in the design space, as shown before. Using only 6 points, this sampling produced the best solution that had been found previously using 15 points.

These results show that HIMO is practical and can result in significant benefit. Although it is desirable that the response surface is sizing-optimal so that the topology optimization is more meaningful, sizing optimization may take a long time to run for large FEA models. In those cases, the feasible surface could be used to take care of the major performance constraints. After higher level optimization is performed on the feasible surface, the topology-optimized solution can go through sizing optimization to find the final sizing-optimized and topology- optimized design. Of course, performing the optimization using optimal surfaces may result in optimal solutions, whereas performing the optimization using feasible surfaces may result in sub-optimal solutions. More tests of sampling, model selection, model building, and optimization is necessary to determine statistically meaningful conclusions regarding how the new different sampling methods impact the optimization results.

4.2.   Pilot Numerical Tests with ESO

The same pilot problem as in the last section was also solved with evolutionary structural optimization approach, to compare HIMO with ESO and to confirm the solution found by HIMO.   ESO is based on the concept of slowly removing inefficient materials from a structure so that the remaining structure evolves towards the optimum. The lowly stressed area is assumed to be under-utilized and will be removed subsequently. By gradually removing material with lower stress, the stress level becomes more and more uniform throughout the new structure.

The procedure is as follows: First, the design domain is discretized as a finite element mesh. Loads and boundary conditions are then applied. Finite element code is run, and the stress is found. Usually, low stress areas are shown that are below threshold. The elements associated with these areas are eliminated. Von Mises stress is frequently used for isotropic materials.

The cycle of finite element analysis and element removal is repeated until a desired optimum is reached, e.g. no stress is below 25% of the maximum stress in the structure. The result may be not the best solution, but the process provides many improved designs. The shape and/or topology at each steady state may be chosen as the final design.

In the application of basic ESO, typical problems are 2D, with each element's stress state being checked. In this application, the problem is a 3D case. Thus, solid elements are needed. The model is presented in Figure 9A. The channel portion is shown in Figure 9B, with the inside filled with solid elements.

59

To be more efficient and effective, variable linking is needed. Since the goal is to find the optimal depth of the reinforcing channels, each layer of the solid elements, with its mid-plane parallel to the end sheet or Y-Z plane, is placed in the same group. Instead of checking the stress in each element, the maximum stress in each group is compared with the maximum in all other groups. The minimum of these maximum values among all the groups represent the group or layer that is not used efficiently or can be deleted.

After the first run, the fourth groups from the end sheet in both channel regions show the minimum stress among all the groups in each channel region respectively. Thus, both groups are deleted. The new model and a highlighted channel region are shown in Figure 10. After the second run, the sixth groups in both channel regions are deleted. Similar actions are taken after additional runs. Finally, after the sixth run, only the seventh groups are left together with the groups for the end sheet portion and for the eighth group, the channel bottom group. The stresses in both of the seventh groups show the minimum stresses in the channels, so they are deleted. The result for the channel region is shown in Figure 11. Only the sheet portions and the bottom layers are left. The stress output for each of the seven runs is shown in Table 3. ESO has led to the same conclusion as HIMO: The deepest channels are the most efficient (optimal) solution for this assembly.

Table 3. The maximum stress output in each group after each running

(von Mises stress) (ksi)

(G: Groups; R: run; C: Channel)

G	R 1 C1	R 1 C2	R 2 C1	R 2 C2	R 3 C1	R 3 C2	R 4 C1	R 4 C2	R 5 C1	R 5 C2	R 6 C1	R 6 C2	R 7 C1	R 7 C2
1	15	14	16	15	16	14	15	14	17	18	19	18	19	17
2	6.0	5.6	7.5	7.0	7.4	6.9	7.2	6.7	8.0					
3	3.3	3.2	7.2	7.1	7.2	7.0	6.9	6.8		8.4				
4	2.6	2.6												
5	3.8	3.8	6.6	6.6	6.0	5.9								
6	5.0	5.0	5.3	5.3										
7	6.4	6.3	6.7	6.6	7.9	7.7	8.8	8.6	9.0	8.7	9.1	8.9		
8	7.7	7.5	8.2	7.9	9.5	9.2	11	10	11	11	11	11	17	16

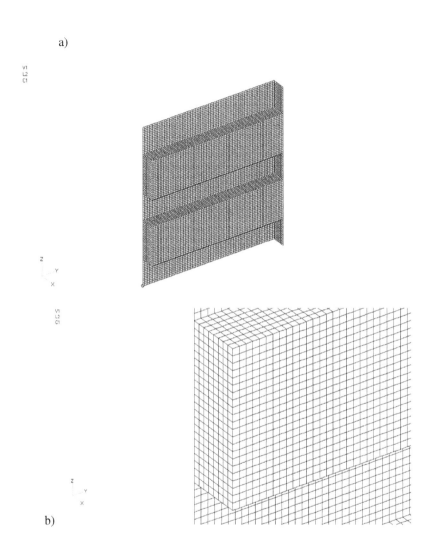

Figure 9. Pilot test with ESO: Model (a: whole model; b: local view of the channel region)

a)

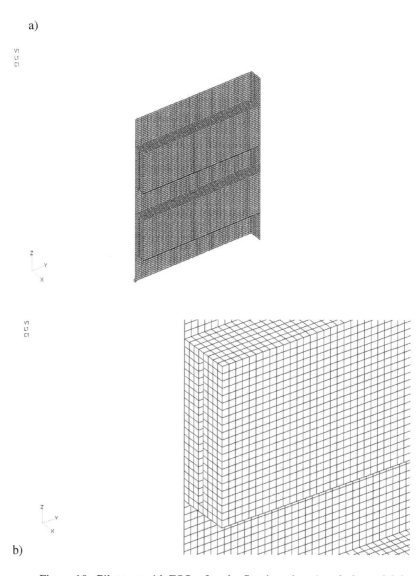

b)

Figure 10. Pilot test with ESO, after the first iteration (a: whole model; b:
local view of the channel region)

a)

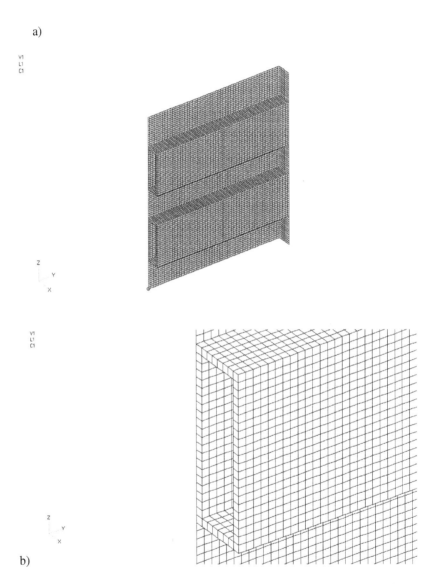

Figure 11. Pilot test with ESO, after the optimization (a: whole model; b: local
view of the channel region)

4.3. Real Applications

HIMO was used to optimize an end structure layout, which is the whole subassembly from which the bottom portion used in the pilot tests was extracted. The existing assembly has an end sheet, 7 channels, plus a corner post for reinforcement. From previous experience and knowledge, channels are very efficient when compared with many other shapes for the similar loading conditions. So it was decided to keep using channels for reinforcement. The corner post also serves other purposes beside reinforcement, so it was kept also.

The layout questions include how many channels should be used, what channel depths and heights should be used, and where the channels should be located. The sizing variables are the thickness of the end sheet, the channels and the post. The goal is to minimize the assembly weight while satisfying the stress and buckling constraints as well as minimum thickness and discrete thickness values corresponding to the sheet metal gages. The topology constraint is the sum of the channel heights and spacing between them plus the top and bottom spans is equal to the space allowed. The end structure assembly is loaded with pressure, uniformly distributed on the end sheet with two load cases.

As for the boundary conditions, since HIMO needs many small models to quickly and reliably evaluate many options, the whole vehicle model was compared with a small model that has only the end sheet, the channels, and the corner post. For the latter, as in the pilot test, the borders are pinned without any displacement, but can rotate. Only half of the structure needs to be modeled because of the symmetry of the structure, loading, and boundary condition. Therefore, a symmetric restriction was applied to the centerline in order to simulate the other half. Later, for buckling analysis, the whole end structure was modeled to guard against possible non-symmetric modes. The initial comparison of the displacement and stress output between the accurate model of the whole structure and the end-only model showed very similar results. But when buckling results were compared, the results from the simplified model were inaccurate, and could not even begin to serve as rough estimates for the real behavior determined using a more complete model. After several trials, it was found that the model representing "one quarter of one quarter" of the whole structure provided very accurate results. The time for one model run was reduced to about ten minutes from more than four hours to run a full "one quarter" model (on a 600 MHz computer).

Regarding problem formulation, a previous similar project with lower loading had resulted in five channels (using HIMO). Although the load values were much higher for the current car project, the author suspected that five channels could still meet the new load requirements. Also from the previous project, the author learned that when the channel depths are allowed to reach the highest limit, the stress and buckling eigenvalues can be reduced considerably, with only a slight increase in weight. So the depth would be the upper limit. Finally, five variables representing five channel heights and five spacing variables remained, resulting in a total of ten topological variables.

Once the problem was formulated, the first task was to explore the whole topology design space to locate the feasible subspace where the designs can satisfy the buckling constraint without changing the designs that can satisfy the stress constraint and keep the weight low. Using the new sampling method (LHMDmd) that finds the sample group with the lowest C criterion value (see Chapter 2, p39) among one thousand or more groups of Latin hypercube samples, a group of twenty samples were designed that met the topology constraint requirement. The intention was, to save the effort for building and optimizing the metamodels, to try only ten samples to quickly explore the whole space. If they turned out to be feasible and the response surface needed to be built, the other ten samples could be tested, resulting in twenty sampling points for the surface. On the other hand, if some designs are far away from the limits, then those infeasible regions will be ignored before the metamodel is built. More sampling will be conducted for the feasible region and the metamodel is only for the feasible region.

Ten tests were done. Without changing the plate thickness of the end sheet and the channels from the sizing optimization outputs, the models were run to find out buckling eigenvalues. Some were far below the limit, whereas others were close. The sampling plan or topology designs were sorted according to the eigenvalues. The trend became clear that when the channel height is increased (within some limits) and the spacing between the channels is reduced, the eigenvalues get higher. The possible feasible subspace was set by specifying the bounds for the topology variables. Another group of twenty samples was designed using LHMDmd. Test results showed that the subspace was still not small enough, since the eigenvalues were still not close enough to the limits. Further reduction of the subspace was made. New samples were designed and tested. After several more trial samplings, the near-feasible subspace was found. At this point, the plate thickness either does not need increase or needs to increase only slightly in order to meet the buckling resistance requirement. Twenty samples were planned and tested. All resulted in feasible designs in terms of the stress and eigenvalue constraints and optimal designs for sizing variables.

The Kriging surface with the zero-order or constant regression polynomial model and the Gaussian correlation model was then fit to the 20 points. The cross-validation shows that the root mean square error (RMSE) is 99.68 N (22.41 pounds), the maximum error is 204.71 N (46.02 pounds), and the relative RMSE (RMSE divided by the mean of the weight response) is 2.47%. A genetic algorithm followed by SQP was then used to find optimal topology.

Since the relative root mean squared error was still not very small, a more promising subspace was sought to further confine the region to be explored. Twenty more samples were tested and then sorted by the weight of the assembly. The subspace defined by the designs with the ten lowest weight values was used as the promising subspace for the next sampling group.

Ten more samples were designed and tested. Together with the ten best samples in last group, twenty samples were used to build the Kriging surface with constant regression model and Gaussian correlation model. The cross-validation results are the following: RMSE is 46.04 N (10.35 pounds), the maximum error is 126.77 N (28.5 pounds), and the relative RMSE is 1.16%. The genetic algorithm followed by SQP found a solution that is quite close to the best one tested in the newest group, which is also the best among all the options tested. Since this new optimal result is not significantly better than the last one, only about 30 pounds lower in weight, and since much better solutions were not expected based on experience, the optimization process was stopped.

The final design was fabricated based on design that results from the final sizing-optimization after topological optimization. The weight was reduced by over 2000 pounds, or about 36 %. The new design also met several other functional and manufacturing requirements and overload concerns (otherwise the weight could be even lower). Using five channels instead of seven considerably reduces the company's manufacturing cost.

This chapter has presented the test results of using HIMO for real applications, showing its applicability and effectiveness for real design of 3D structural assemblies. The pilot test was also conducted by using ESO with the same conclusions for the design. HIMO was also applied to real design optimization for a car end structure, resulting in 18% and 36% weight savings for two projects (two different car designs) and significant cost savings for manufacturing. The pilot tests and real application are summarized in the Table 4.

Although similar procedures can be used for applying ESO to other similar problems, it may be difficult to use ESO for many types of problems. For instance, when using ESO, solid elements are needed to fill the design space based on the ground structure approach, which involves long stress analysis runs, plus considerable time for pre-processing and post- processing. Although for the particular problem (the pilot test), ESO can be used to evaluate the options by applying variable linking, HIMO is much more versatile for general 3D structures, especially for large-scale problems. For instance, it is very difficult to use ESO for the five-channel problem presented above. For another example, in boxcar design, the positions of the lateral support in the underframe are an important layout problem. HIMO can be easily applied, but it would be very difficult to use ESO to fill the whole design space with material and then to find the optimal solution.

Table 4. Pilot tests with HIMO and ESO and real application with HIMO

Tests	Number of design variables	Topology design	Sizing design (thickness)	Iterative and interactive ?
Pilot with HIMO	2	Deepest channels	optimal	yes
Pilot with ESO	14 (by applying partial ground structure approach)	Deepest channels	Needs sizing optimizer to find the optimal	yes
Real design with HIMO	10	Optimal channel height and spacing between channels	optimal	yes

5. Test results for SOTO — using a sizing-optimizer to optimize topology and partial ground structure approach

5.1. Numerical Tests

SOTO was applied to several classical 2D topology optimization problems. The objective is to minimize the weight. The constraints are the stress or displacement limits. The design variables are the plate thickness. The results are similar to those by the homogenization method and ESO. Three of the numerical tests are reported here.

The first is a clamped deep beam with a single load 3559 N (800 lb.) at the center of the right side. The Young's modulus is 100 GPa and Poisson's ratio is 0.3. The stress limit is 165 MPa (24000 Psi). The width is 254 mm (10 in.) and the height is 610 mm (24 in.). There are 800 quadrilateral elements. The desired variable or thickness bounds are 1 mm (0.04 in.) and 5 mm (0.2 in.). The FEA model and iteration history (after iteration 1 and the last) and the final design are shown in Figure 12. The last plot is one of the options for the final design, based on topology optimization result and was size-optimized. The initial thickness is 2.54 mm (0.1 in.) and final one is 1.4 mm.

The initial model and iteration history by applying homogenization method and ESO can be found from the similar figure in Hassani and Hinton, 1999 (no thickness is provided) and the similar figure in Xie, et al, 2002 (the thickness is 1 mm).

Figure 12. Example 1 Iteration history and final design by SOTO

The second test is a Michell type structure. The first truss solutions of least weight and a general theory for solving the problems were published by Michell (1904). Many of his solutions have been used as benchmark problems by later researchers in structural optimization.

The design space is 10 m by 5 m. The two bottom corners have the support: the left is pinned restricting movement in both directions; the right is simply supported. The Young's modulus is 100 Gpa and Poisson's ration is 0.3. The load is 1000 N. The stress limits are -50ksi and 50 ksi. The iteration history and the final solution are shown in Figures 13 and 14. The final solution is based on the result from the fifteenth iteration and smoothed for manufacturing considerations.

The initial model and iteration history by ESO can be found from Xie, et al, 2002.

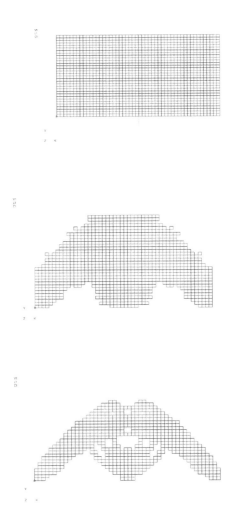

Figure 13. Iteration history of Test problem2 by SOTO

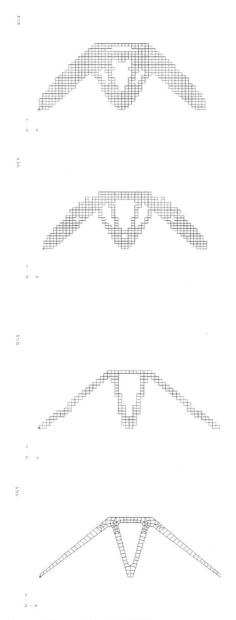

Figure 14. Iteration history of Test problem2 by SOTO

The third test is another Michell type structure, as shown in Xie, et al, 2002. The support is around the circle. A single load 50 kN is at the center of the right side. The Young's modulus is 205 GPa and Poisson's ratio is 0.3. The displacement limit is 9 mm vertically at the loading point. The rectangular space is 0.55 m by 0.4 m. There are 2800 elements. The desired thickness bounds and the control limits, are 1 mm and 5 mm. The FEA model and iteration history (after iteration 1, 2, 3, 6, 7, and the last) and the final design using SOTO are shown in Figure 15. After the first iteration, more than half of the elements were deleted. Iteration 6 took a long time to run. Many chains in the diagonal portion were suspected to have caused the problem. The portion was re-meshed to eliminate the chains. Iteration 7 took about 1 minute to run. The last plot is one of the options for the final design, based on the topology optimization result, and was size-optimized. The final thickness is about 2 mm. (0.065 in.). The figure shows the iteration history and final design. It can be seen that the last iteration result and the final design are very close to the theoretical solution. The similar figure in Xie, et al, 2002, shows the problem and iteration when using ESO.

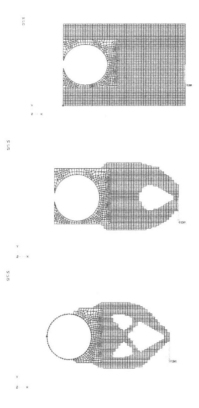

Figure 15 (1). The initial model and iteration history by SOTO, top to bottom

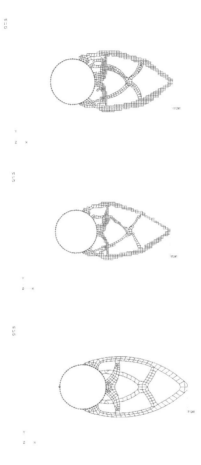

Figure 15 (2). The initial model and iteration history by SOTO, top to bottom

5.2.   The pilot test of SOTO on a real application

To see the applicability and the potential of SOTO, a pilot test was done to analyze part of a real structure. This simple structure is the bottom section of an end assembly of a vehicle, consisting of an end sheet reinforced by two channels and a corner post. For finite element analysis, only half is modeled because of the symmetry of the structure, loading, and constraints. The FEA model (for plot only) is shown in Figure 16. The border is pinned so that all the translation movement is restricted, except the symmetric center where the symmetric constraint is applied to simulate the other half.

The objective function is the weight of the structure. The layout question is to determine the two channel depths that result in the lowest weight for the feasible structure that meets the strength requirement.

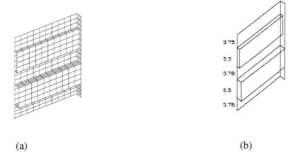

(a)   (b)

Figure 16. FEA model used for pilot tests of SOTO (a: the model; b: dimension)

The half model dimensions (in inches) are the following: end sheet height, 22.25; width, 20; corner post width, 2; channel height, 5.5; spacing between channels and top/bottom spans, 3.75. The stress limits are -50 ksi, 50 ksi. The thickness bounds are 0.1046 and 0.5 inches. Young's modulus is 29000 ksi, and Poisson ratio is 0.3.

If one used a ground structure approach and distributed density problem formulation, solid elements would be used to fill the space within the channels, or between the channel bottom plates and the end sheet. Each element would be assigned a unique density. For applying SOTO, different problem formulation, variable linking, and partial ground structure were sought. The actual assembly is made from metal sheets, so plate elements are more appropriate. Element thickness, not the density, is used as the design variable. This is one of the choices available in the sizing optimization module. MSC/Visual NASTRAN for Windows was used for the pilot tests and all the other tests reported in this dissertation.

Secondly, if plate elements are used and ground structure approach were followed, the plates should fill the space within the channels. For instance, a 3-D grid structure could be used, comprising plates parallel with the XY plane, YZ plane and XZ plane. Since the purpose in this problem is to find the best channel depths, only the plates parallel to the channel bottom plate are needed, which are parallel to the YZ plane. Thus, the approach is referred to as a partial ground structure approach. Seven plates were created in each channel, parallel to the channel bottom, shown in Figure 17. They are all called bottom plates. Thirdly, since each plate or channel has one thickness, it makes sense that the each bottom plate as whole, rather than every element in the plate, is assigned a design variable. Consequently, variable linking is used.

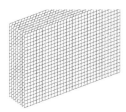

Figure 17. Arrangement of parallel plates in the channel region (Pilot test of SOTO)

Associated with the bottom plates are the channel legs, i.e., the top and bottom plates. Each channel has two legs, shown as the two plates parallel to the XY plane. Not shown in Figure 17, each channel has nine rows of elements along Y direction in either top or bottom leg. Should the optimizer "select" the lower depth, the row in the upper X direction might be assigned "zero" or small thickness. Hence, the elements in each row should be the same and assigned a unique design variable. So the variable linking is again used. Next, the variable bounds are decided. The variables are allowed to be zero, so the "practical zero" of 10^{-12} is used as the lower bound to avoid possible numerical problems (this is a much lower value than used for the 2D problems). The upper bound is determined by the thickness upper bound of 9.5 mm (3/8 in.) used for the real product. The stress limit is 345MPa (50000 Psi).

In summary, the optimization problem is as follows: Find the optimal depths for the two channels, to minimize the structure weight, subject to the stress constraints. Specifically, there are eight design variables associated with the bottom plates and eight design variables associated with each of the channel legs. The corresponding rows with the same X value in the top and bottom legs share the same design variable.

After the first iteration, two plates in each channel closest to the end sheet were assigned the values in the order of 10^{-4}, and the next two plates in each channel were in the order of 10^{-3}. All the other plates were in the order of 10^{-2}. After the bottom plates in 10^{-4} were deleted, the second iteration was started. After finishing the run, the two plates in each channel closest to the end sheet were in the order of 10^{-4} and 10^{-3} respectively, and all the other plates were in the order of 10^{-2}. The third iteration was started after the four plates in 10^{-4} and 10^{-3} were deleted. After this run, the last two plates in each channel had value of 3.5 and 5.3 respectively. The two thinner plates closer to the end sheet were deleted. At this point, only one plate, the one that was farthest from the end sheet remained for each channel. The result is that the deepest channel is the best. This was obtained by following the same procedure as used in the 2-D examples, i.e. gradually deleting the plates with smaller thickness leads to the improved design.

The result is consistent with the intuitive thinking. A little deeper channel results in a little weight increase, but the stress is reduced much (or buckling eigenvalues can be significantly higher if buckling is another constraint), which is very beneficial. This result was also found with HIMO as reported in Chapter 4.

In this case, "the winner takes all" strategy employed by SOTO results in the optimal solution. The thickest plates remain and all the other thinner plates are gone. More research is needed to determine if this result is problem-dependent or not.

In retrospect, the best topology was already clear immediately after the first iteration: the farthest bottom plate had the largest thickness among all the parallel bottom plates. It was the same for each channel. This indicates that the subsequent work was not needed, at least in this case, but served instead to confirm the results.

5.3. Real Application of SOTO

Next, SOTO was used to optimize the full end structure layout, which is the whole assembly from which the bottom portion was used in the pilot test. The end structure assembly is loaded with pressure, uniformly distributed on the end sheet with two load cases. HIMO was applied to this same problem, as discussed in Chapter 4. It was assumed that the deepest channels would result in the best design for the different depths. The best design was found to have five channels.

Assuming the topology design from HIMO is good, the purpose of applying SOTO for this problem is to confirm that the deepest channel is the best. By keeping the final topology design resulted from HIMO, nine more bottom plates were inserted within each channel of the five channels, again using partial ground structure approach. They are uniformly distributed and parallel to each other. As done in the pilot test, each bottom plate thickness was a design variable. Further, the thickness of each row of the elements in each leg is represented as another group of design variables. At the same X coordinate value, the top and bottom rows share the same design variable. So there are 5 x 10 bottom plate thickness variables and 5 x 10 channel leg plate thickness variables with each representing two rows of the elements, for a total of 100 design variables.

The optimization problem for this real design project was to find the optimal depths for the five channels, in order to minimize the end structure weight, subject to the stress constraints. Following the same process used in the pilot test, bottom plates with the smallest thickness were gradually deleted over several iterations. Finally, for every channel, the farthest bottom plate remained, the winner taking all again. Thus, the conclusion is that the deepest channel within the limit is the best in terms of the weight, subject to the stress constraint.

Again, in retrospect, the best topology was already clear after the first iteration: the farthest plate for each channel has the largest thickness among all the bottom plates within the same channel. The subsequent iterations were not needed for this case, but served instead to confirm the result.

In addition to finding the optimal depths for each channel, SOTO was also used to seek the solution for the more general layout problem including the following questions, as an alternative to using HIMO: What shapes should reinforcements be: channels, I-beams, T-beams, angles, or some other shape yet to be determined? If channels are shown to be better, how many channels should be used? What channel depths and heights should be used? Where should the channels be located? The goal is to minimize the subassembly weight while satisfying the stress and buckling constraints as well as the minimum thickness values and the discrete thickness values corresponding to the sheet metal gages. The topology constraint is the sum of the channel heights and spacing between them plus the top and bottom spans must be equal to the space allowed.

An initial attempt was made to seek the answers to the general questions, except the question of depth (assuming instead that the deepest is the best). Again, partial ground structure approach was used. There are many uniformly distributed "legs" and one sheet as the "bottom" parallel to the end sheet, covering the whole plate space from the top to the bottom. Again, variable linking was used. All the elements in a "leg" share the same thickness that is a design variable, and all the elements in the same row at the same height in the bottom plate share one thickness that is another design variable. There are 321 legs and 320 rows, thus 641 design variables are associated with the legs and bottom sheet. Initially,   the sizing optimizer with the stress constraints assigned very small thickness to many "legs", essentially "pushing" the material to the bottom plate. But when buckling was checked, the eigenvalues were too small to meet the stability requirement. Unfortunately, the optimizer used does not have the buckling constraint option. The answers to these questions will have to be deferred.

All the tests shown in this chapter are summarized in the following Table 5.

In this chapter, the tests using SOTO indicate that SOTO can be used to solve the same numerical problems solved by homogenization methods and ESO. SOTO was also successfully applied to real design optimization, showing its applicability for real engineering problems. Since SOTO uses the sizing

85

optimizer, it is restricted by the capability of the optimizer. For instance, it is mentioned if the sizing optimizer does not have the capability for buckling constraint, SOTO cannot be used for the problem. Another problem associated with the optimizer also restricts SOTO: the optimizer could become trapped in a local optima. There are strategies such as assigning different initial plate thickness that can help the optimizer to move out of such traps in some cases. Another issue is related to time. Optimization takes several analysis runs, thus it takes longer to finish one optimization operation than would be required for just analysis run. For simple problems, ESO might be able to improve the design more quickly than SOTO, especially if the partial ground structure approach is applied. However, as mentioned before, since ESO does not optimize the thickness, the wrong selection of the initial thickness might result in a less optimal topology design. If many trials of the initial thickness are needed, it might take ESO a longer time to find the solutions reached more directly by SOTO.

Table 5. Test summary for Chapter 5

Tests	Topology design	Sizing design (thickness)	Iterative and interactive ?
Numerical 1 with SOTO	Similar to those by Homoge. and ESO	optimal	yes
Numerical 1 with Homogenization	Similar to those by SOTO and ESO	Needs sizing optimizer to find the optimal	No (one shot approach for simple prob.)
Numerical 1 with ESO	Similar to those by Homoge. and SOTO	Needs sizing optimizer to find the optimal	yes
Numerical 2 with SOTO	Similar to that by ESO	Optimal	yes
Numerical 2 with ESO	Similar to that by SOTO	Needs sizing optimizer to find the optimal	yes
Numerical 3 with SOTO	Similar to that by ESO	Optimal	yes
Numerical 3 with ESO	Similar to that by SOTO	Needs sizing optimizer to find the optimal	yes
Pilot with SOTO	Similar to those by HIMO and ESO	Optimal	yes
Real design with SOTO	Optimal: deepest channels	Optimal	yes

6. Exploration of some issues for experimental designs for computer experiments

6.1. Comparison of 18 design types in terms of prediction accuracy

Of the 18 design types, four types were thoroughly tested, as described in Section 6.1.1. In Section 6.1.2, all 18 design types were tested and compared.

6.1.1. Comparison of four primary experimental designs

Random sampling (Rd), Latin hypercube design (LHD), maximin Latin hypercube design (LHMm), and minimized centered L_2 discrepancy Latin hypercube design (LHCL2) are compared using twenty test functions in 2D cases, and five functions for 5D and 10D cases. For each n (dimension) and m (sample size) values, four experimental designs were used for sampling; Kriging models were built to approximate the function; errors in the models' response were found by comparing approximated function values with the actual function values at 10000 verification points within the design space. The verification points were determined using LHD. The points are exactly the same for every design type compared. Table 6 summarizes the results of these tests. More details are provided in the Appendix.

The test results indicate that there is no significant difference in the means of RMSE (root mean square error) or maximum errors in many cases. When there are significant differences, LHD is nearly always better than random sampling; LHMm is better than other designs in 2D cases when the sample size is large (18-22), often worse than LHD and LHCL2 in 5D cases, and nearly always worse than LHD in 10D cases; LHCL2 is often better than LHD in 2D cases.

Table 6. Summary of the tests comparing RandSamp, LHD, LHMm, LHCL2
(The percentage of the total cases that have significant differences among different designs)

Comparing 4 design types Rd, LHD, LHMm, LHCL2					
1)	Compare RD LHD LHMm LHCL2				
			N=2	N=5	N=10
			200 cases	50 cases	50 cases
			Percentage, %	Percentage, %	Percentage, %
	p-v <0.01		43.5	14	26
	p-v <0.05		54	24	42
2)	Compare Rd LHD				
	p-v <0.01		8.5	2	4
		Rd better	0.5	0	0
		LHD better	8	2	4
	p-v <0.05		24	2	14
		Rd better	1	0	2
		LHD better	23	2	12
3)	Compare LHD LHMm				
	p-v <0.01		27	10	28
		LHD better	13.5	10	28
		LHMm better	13.5	0	0
	p-v <0.05		42.5	22	48
		LHD better	25	22	46
		LHMm better	17.5	0	2
4)	Compare LHD LHCL2				
	p-v <0.01		8.5	8	2
		LHD better	1	6	0
		LHCL2 better	7.5	2	2
	p-v <0.05		15.5	20	12
		LHD better	1.5	12	10
		LHCL2 better	14	8	2
5)	Compare LHMm LHCL2				
	p-v <0.01		39	12	30
		LHMm better	15	2	2
		LHCL2 better	24	10	28
	p-v <0.05		51.5	28	46
		LHMm better	21.5	6	4
		LHCL2 better	30	22	42

Explanation:
Design types compared: Rd (random sampling), LHD (Latin hypercube design), LHMm (Maximin LHD), LHCL2 (minimized center discrepancy LHD).
N: dimensions.

Percentage: the number of the cases where the different designs are not equivalent (at the given p-value) divided by the total number of cases in the group.
p-v: probability value from ANOVA (one-way).

6.1.2.   Comparison among 18 design types

Eighteen design types, including uniform designs (UD), are compared with one-way and two-way ANOVA. In 2D cases, out of total 100 cases, 37 cases or 37% show significant differences (p-value < 0.01) in the means of the errors generated by different designs; 60 cases show p-values below 0.05. As an example, the means of RMSE (means of the 10 groups with the same dimensions and sample size) are shown below (Figure 18) for Axis parallel hyper-ellipsoid function. Table 7 summarizes the numbers with p-values below 0.01 and 0.05.

Table 7. The percentage of the cases that have p-values below 0.01 or 0.05

	N=2	N=5
	100 cases	50 cases
	%	%
p-values		
p-v<0.01	37	16
p-v<0.05	60	26

Note: The tests are whether or not there is any statistically significant difference amongst all of the 16 design types.
N: number of dimensions.
The number of cases: the total number of the cases in the group.

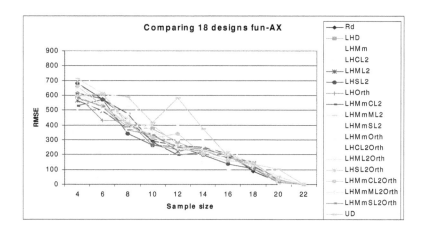

Figure 18: Comparison of 18 designs in 2D cases.

6.2. Could enough samples be more important than better designs?

In the cases tested, there is often no significant difference between the mean errors generated by the different design types, even when the critical p-value is selected as 5%. In the 2D cases where there are significant differences, it is likely that the increase in accuracy resulting from adding more runs, will be greater than that achieved by using "better designs."

Figure 19 shows the comparison of the RMSE for one test function (Branins' rcos) when four design types (random sampling, LHD, LHMm, and LHCL2) are compared. Each data point in the figure is the average of 20 output values of the RMSE. It is clear that the sample sizes have much greater impact on RMSE than the design types.

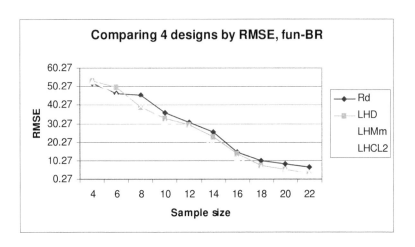

Figure 19. Relative RMSE changes vs. the sample size for Branins's rcos

function when Rd, LHD, LHMm, and LHCL2 are compared in 2D cases.

The examination of the two-way ANOVA tables (shown in the Appendix) for the case above reveal that the portion of the variation in terms of the sum of the squares caused by different sample size is much larger than that caused by different designs. The p-values for the two-way ANOVA for this case are: [0.0001, 0, 0.2053]. The first number shows design type effect, the second sample size effect, and the third interaction between the two. Since the first value is far below 0.01, the different design types do impact the accuracy. Even so, the sample size does have greater impact, not to mention the cases when there is no significant difference among the mean errors generated by different designs.

The following pie charts in Figure 20 are generated from the ANOVA tables based on testing the first ten functions, comparing Random Sampling, LHD, LHMm, and LHCL2; with n = 2, 5, and 10. This figure shows clearly that the run number effect (red color) occupies a much larger portion of the total variation than the design type effect (blue color). The yellow color shows the portion due to interaction and within group variation.

For most of the cases tested, it appears that enough samples may be more important than "better designs" when Kriging models are used for similar dimensions. In many cases, adding a few more samples is likely to increase the accuracy of the approximation more than by using "better designs." Using enough samples might be more efficient than spending time to try to find better designs in many cases.

The sample size determination for metamodeling is likely to be related to the following factors as a starting point: the number of design variables; the characteristics of the approximation model; the number of parameters estimated for the model; the complexity of the underlying function (relationship); the radius of the hyper-sphere containing the design space; variable conversion considerations, normalization, the units used, related to the radius; desired prediction accuracy; and experimental design approach.

Many values in the third column of the p-values for two-way ANOVA are below 0.01, indicating there is significant interaction between sample sizes and design types in many cases.

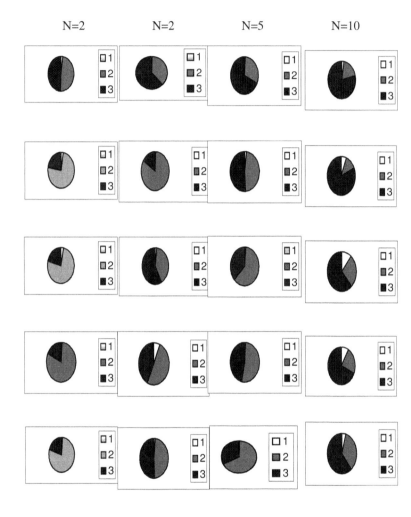

Figure 20. The portions of the variation caused by designs, sample sizes, interaction plus within group variation

1: The portion of the total variation caused by different design types

2: The portion of the total variation caused by different run numbers,
   Increment: 2 (for n=2); 5 (for n=5, 10).

3: The portion of the total variation caused by interaction between columns and rows, or designs and run numbers, plus within group variation.

6.3. Could more uniformly distributed sampling be always better for computer experiments?

It has been shown that LHD and LHCL2 are often better than LHMm when there are significant differences between the means of the errors. The following figure shows one example of the average RMSE resulted from using LHMm and LHCL2. The test function is De Jong's function 1. The number of design variables is 2.

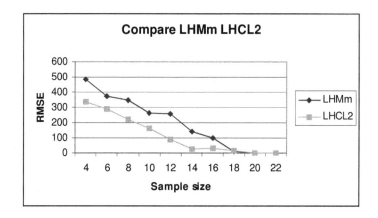

Figure 21. Comparison between LHMm and LHCL2 based on RMSE for De Jong's function 1.

However, according to the simulation results, LHMm almost always generates more uniformly distributed samples in two-dimensional designs, comparing to the samples generated by LHCL2. It is generally agreed that more uniformly distributed sampling is preferred (see e.g. Santner, et al, 2003, p124). Why do the results here appear to contradict this feeling? For further investigation, many groups of ten samplings from LHMm and LHCL2 each were tested. For each sampling, the twenty test functions were approximated with Kriging models. The error used for comparison is the relative root mean square error (this is unique for this group of tests) as the follow:

$$RMSE = \sqrt{\frac{1}{5000} \sum_{i=1}^{5000} \left(\frac{Ft(i) - Fa(i)}{|Ft(i)| + \varepsilon} \right)^2}$$ (46)

Ft: True response; Fa: approximation; ε: 10^{-6}: to guard against possible Ft = 0.

For verification, only one sampling group of 5000 points was generated and used by all of the tests. One way ANOVA was applied to compare LHMm and LHCL2. This process was repeated ten times for each combination of dimension (n=2) and sample size (m=4, 6 , ..., 22). The sampling plots, ANOVA plots together with the p-values for two cases are shown in Figure 22. All the results are shown in Appendix.

In most cases, there is no significant difference between the means of the relative errors generated by the two different designs, though in many cases it is clear that LHMm generates a more uniformly distributed sampling. However, LHMm is significantly worse than LHCL2 instead of being better.

More cases were similarly tested: 1) comparing LHMm with LHD, and 2) comparing LHMm with Random Sampling. Again, LHMm generates more uniformly distributed sampling, yet no significant differences in the mean of the errors are seen. In some cases, LHMm is worse than the other design despite its more uniformly distributed sampling.

n=2, m=10

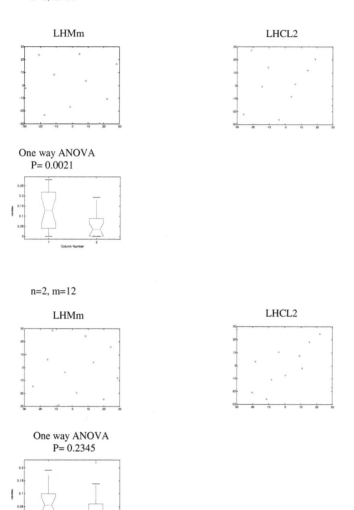

Figure 22. Sampling plots and one-way ANOVA plots: comparing RMSE resulted from sampling with LHMm and LHCL2
n: number of dimensions
m: sample size

7. Discussion and conclusions

7.1. HIMO and SOTO

This research has developed two new systematic approaches for addressing difficult structural topology optimization: HIMO and SOTO. The key features of HIMO are as follows.

1) All of the performance constraints are dealt with at the sizing optimization level only, thus avoiding the stress constraint difficulty for topology optimization.

2) The number of topology design variables is therefore much smaller than required by many other topology optimization approaches, which greatly reduces the effort needed to optimize the topology.

3) The sizing optimizer is used to find the feasible and optimal design among all the sizing options nested within one topology option, for every sampled point.

4) The response surface of the objective of interest (e.g. weight) is an optimal surface in terms of sizing variable or plate thickness in continuum cases, and topology optimization then works on this surface to find best topology designs.

Both the pilot numerical tests and the real applications show that this hierarchical, interactive, and metamodel based optimization approach (HIMO) works well for difficult layout optimization problems involving real structures. HIMO is also very flexible. When the budget is tight, and less sampling and testing can be done, a significantly improved design is still very likely to be found. When the budget allows more work, and finding the optimal result is very important, more sampling and testing can be conducted, resulting in much better or optimal solutions.

It may be desirable to incorporate HIMO into FEA or structural optimization programs, in order to provide automatic HIMO for topology optimization of structural systems. The programs would need to include modules for design and analysis of computer experiments (DACE). DACE includes tools for experimental design, model selection and building, and global optimization (e.g. genetic algorithms combined with mathematical programming methodologies). In preprocessing, the programs would need to quickly generate many different models or different topology designs according to the sampling plan. Incorporating parametric modeling used in 3D CAD programs might be helpful. The solvers would need to automatically do batch or parallel processing and to automatically generate reports on the objective function values e.g. weight. In post-processing, the programs would need to build, verify, and optimize the metamodels.

For SOTO, the numerical tests, the pilot test, and the real application show that the new approach using a sizing optimizer to optimize topology can work for many cases. It meets the need to automatically search for the optimal plate thickness as well as topology, since the plate thickness may impact the topology optimization results. This interaction is a problem for the popular ESO method discussed in the literature. SOTO, therefore, offers a second heuristic and systematic search method for addressing difficult topology optimization problems of real structures.

Both HIMO and SOTO were also compared with ESO using several benchmark problems. They were also compared to the homogenization method for one particular case. The final results are quite close for each of these problems. A qualitative comparison of several methods is discussed below.

To search for the best solution using brute force and exhaustive search, for a problem with 10 variables and 10 levels for each variable, 10^{10} options would need to be evaluated or sampled. This is clearly infeasible. Even for smaller problems, exhaustive search is very inefficient, limiting its use to very small and simple structures. One benefit of exhaustive search is that a global optimum can be found.

Engineers often use trial and error in the design process. Since it lacks a systematic search, this method can be slow and may fail to locate the optimal design. It can, however, successfully improve designs.

Homogenization-based approaches are the primary mathematical methodologies for topology optimization. Major developments have occurred in this area, including the capability to address stress constraints in some cases. However, they are not yet able to address complicated 3D structures, especially large-scale structures.

ESO has also been used in many cases. It is usually slow and difficult for use in large-scale structures, and may miss the optimal. Because the initial thickness may be selected wrong, the result of topology optimization may be adversely impacted. Despite these limitations, ESO, like SOTO, can help engineers systematically search for much improved designs.

HIMO offers several advantages over other methods. It can handle stress, displacement, buckling, and other constraints. Because the number of design variables is much smaller than required by other methods, it can handle large scale structures with high speed. Until fully automatic approaches are developed, engineers can be trained to use HIMO to develop improved designs. The result can be optimal for both iterative methods like HIMO and for fully automatic approaches.

SOTO and ESO are heuristic methods. Unlike ESO, when using SOTO the computer helps to optimize the plate thickness for continuum plate structures, which may affect the final result for improving topology and plate thickness. Like ESO, SOTO cannot be counted on to find the optimal result, but it can help engineers to systematically search for improved designs.

Both new methods can be used for topology optimization. When to use HIMO and when to use SOTO depends on the problem to be solved. For complicated problems with many options to be selected and the problem can be formulated as an approximation problem for employing metamodeling, HIMO may be appropriate, such as the channel height and spacing design presented in Chapter 4. For 2D topology design that is not easy to be formulated as an approximation problem, SOTO may be appropriate, such as those numerical problems shown in Chapter 5. Sometimes, both HIMO and SOTO can be used. The selection should be based on which is more efficient. For instance, for the design of the channel depth shown in both Chapters 4 and 5, SOTO may be faster.

Table 8 compares HIMO, SOTO, and several other methods for topology optimization.

Table 8.   Comparison of HIMO and SOTO with Alternative Methods

Methods	Sampling points	Number: topology variables (needed)	Speed	For Large scale?	Stress constr aint	Can the optimal be found?
Brute force	huge number		extremel y slow		yes	close
Trial & error	various		slow		yes	improved
Homogenizat ion-based		Very large	Can be slow	no	devel oping	optimal
ESO		Very large	slow	no	yes	improved
SOTO plus partial ground structure		limited	can be faster than ESO	yes	yes	improved
HIMO	small	Very small	Faster: large-scale	yes	yes	optimal

7.2. Experimental design study

Simulation tests were conducted to determine the approximation accuracy resulting from different experimental design types. For the cases tested, the design types investigated often do not have a large impact on the prediction accuracy, i.e.

there is no significant difference in the means of RMSE (root mean square error) or maximum errors. When there are significant differences, the following are observed. LHD is almost always better than random sampling in 2D cases. In most of the 5D and 10D cases, there is no significant difference between LHD and random sampling. LHMm is mostly better than other designs in 2D cases when the sample size is large (18-20). In these cases, the relative error (RMSE divided by the mean of the response) level is often about 0.01 or lower and therefore it may not be an important difference. LHMm is mostly worse than LHD and LHCL2 in 5D cases, and nearly always worse than LHD in 10D cases. In 2D cases, LHCL2 is mostly better than LHD and often better than LHMm when the sample size is about 14 or smaller.

The uniform designs based on the tables available in the literature (http://www.math.hkbu.edu.hk/UniformDesign/) perform better than other designs for some cases, but perform worse for various other cases.

This is a preliminary investigation based on twenty test functions. Many more functions should be tested. The outputs of computer programs such as finite element analysis might be different from those of the test functions. The numbers of design variables tested were 2, 5, and 10. For large problems, the number of design variables can be much larger. Sample sizes up to 100 were tested. Larger sample sizes may be needed for large problems. Only a small portion of the design types in the literature was tested. Only Kriging models were used. Many other models should be evaluated, such as support vector regression models, models generated by genetic programming, neural network models, radial basis function models, spline models, polynomial models, etc.

Another limitation is that the objective function for combined criteria was a linear combination of the criteria. More sophisticated multi-objective optimization methods are reported in the literature under the heading of multi-objective optimization (MOO) or MDO (multidisciplinary optimization).

Nevertheless, this study reveals that it may be more important to have sufficient samples than to have better designs, at least when Kriging models are used. Detailed guidelines for determining sample size are very much needed and would be invaluable to the users of metamodeling.

It is also shown that LHCL2 is often better than LHMm in terms of prediction accuracy. Yet, LHMm almost always generates more uniformly distributed sampling in 2D. Many tests have shown that more uniformly distributed sampling does not necessarily lead to more accurate models. Thus it is not appropriate to use uniformity as the sole criterion to compare different designs for prediction accuracy. Much more study is needed.

Among the design methods tested, Minimized centered L2 discrepancy Latin hypercube design (LHCL2) or Latin hypercube design (LHD) is recommended for the 2D cases similar to those tested; Latin hypercube design (LHD) is recommended for the 5D or 10D cases similar to those tested, with less sampling effort but better prediction results.

REFERENCES

Allen, T. T., Bernshteyn, M., L. Yu, and K. Kabiri (2003), "A Comparison of Alternative Methods for Constructing Meta-Models for Computer Experiments," The Journal of Quality Technology, 35 (2), 1-17.

Back T., Homeister F. and Schwefel H.-P. (1991), "A survey of evolution strategies", in R.K. Belew and L.B. Booker (Eds.), Proceedings of the 4th International Conference on Genetic Algorithms, Morgan Kaufmann, 2-9.

Barton, R. R., 1994, December 11-14, "Metamodeling: A State of the Art Review," Proceedings of the 1994 Winter Simulation Conference (Tew, J. D., Manivannan, S., et al., eds.), Lake Beuna Vista, FL, IEEE, pp. 237-244.

Barton, R. R., 1998, December 13-16, "Simulation Metamodels," Proceedings of the 1998 Winter Simulation Conference (WSC'98) (Medeiros, D. J., Watson, E. F., et al., eds.), Washington, DC, IEEE, pp. 167-174.

Barton, R. R. 1992. Metamodels for simulation input- output relations, Proceedings of the 1992 Winter Simulation Conference, ed. D. Goldsman, J. Swain, J. Wilson, 41 1-41 8. IEEE, Piscataway, New Jersey.

Bendsoe, M. and Kikuchi, N., Topology and layout optimization of discrete and continuum structures, Structural Optimization: Status and Promise, AIAA, Washington DC, 1993, pp. 517-547.

Bendsoe, M. and Kikuchi, N. 1999: Topology and layout optimization of discrete and continuum structures, Structural Optimization: Status and Promise, AIAA, 517-547.

Booker, A., "Examples of Surrogate Modeling of Computer Simulations," ISSMO/NASA First Internet Conference on Approximations and Fast Reanalysis in Engineering Optimization, 2000.

Booker, A., "Using Metamodels for Engineering Design," INFORMS Seattle Fall 1998 Meeting, Seattle, WA, INFORMS, October 25-28, 1998.

Booker, A. J., Dennis, J. E., Jr., Frank, P. D., Serafini, D. B., Torczon, V. and Trosset, M. W., "A Rigorous Framework for Optimization of Expensive Functions by Surrogates," Structural Optimization, Vol. 17, No. 1, 1999, pp. 1-13.

Booker, A. J., 1998, September 2-4, "Design and Analysis of Computer Experiments," 7th AIAA/USAF/NASA/ISSMO Symposium on Multidisciplinary Analysis & Optimization, St. Louis, MO, AIAA, Vol. 1, pp. 118-128. AIAA-98-4757.

Bulman, S., Sienz, J. and Hinton, E., Comparisons between algorithms for structural topology optimization using a series of benchmark studies, Computers and Structures, Vol. 79, No. 12, 2001, pp. 1203-1218.

Cheng, K.T., Olhoff. N., An investigation concerning optimal design of solid elastic plates, International Journal of Solids & Structures, Vol. 17, 1981, pp. 305-323, in Ref. 2.

Chipperfield, A., Fleming, P., Pohlheim, H., and Fonseca, C. Genetic Algorithm Toolbox: User's guide, http://www.shef.ac.uk/~gaipp/ga-toolbox/manual.pdf

Cohn, M., Dinovitzer, A. 1994: Application of structural optimization. Journal of Structural Engineering. 120, 617-650.

Fang, K. T., 1980, "Experimental Design By Uniform Distribution," Acta Mathematice Applicatae Sinica, Vol. 3, No. 363-372, pp.

Fang, K.-T. and Wang, Y., 1994, Number-theoretic Methods in Statistics, Chapman & Hall, New York.

Fang, K.T., Lin, D.K.J., Winker, P. and Zhang, Y. (2000), Uniform design: Theory and applications, Technometrics, v42, n3, pp237-248

Fang, K.T., Ma, C.X. and Winker, P. (2002): Centered L2-discrepancy of random sampling and Latin hypercube design, and construction of uniform designs, Mathematics of Computation, 71, 275-296.

Eschenauer, H., Olhoff, N. 2001: Topology optimization of continuum structures: a review. Applied Mechanics Review. 54, 331-390.

Goldberg D.E. (1989), Genetic algorithms in search, optimization and machine learning, Addison-Wesley Publishing Co., Inc., Reading, Massachusetts.

Hassani, B., Hinton, E., A review of homogenization and topology optimization I--homogenization theory for media with periodic structures, Computers and Structures, Vol. 69, No. 6, 1998, pp. 707-717.

Hassani, B., Hinton, E., A review of homogenization and topology optimization II--analytical and numerical solution of homogenization equations, Computers and Structures, Vol. 69, No. 6, 1998, pp. 719-738.

Hassani, B., Hinton, E., A review of homogenization and topology optimization III--topology optimization using optimality criteria, Computers and Structures, Vol. 69, No. 6, 1998, pp. 739-756.

Hickernell, F.J. (1998): A generalized discrepancy and quadrature error bound, Mathematics of Computation, 67, 299-322.

Holland J. (1975), Adaptation in natural and artificial systems, University of Michigan Press, Ann Arbor.

Iman, R.L. and Conover, W.J. (1982): A distribution-free approach to inducing rank correlation among input variables, Communications in Statistics , Series B 11,311-334.

Jin, R., Chen W., and Sudjianto, A. (2003): An efficient algorithm for constructing optimal design of computer experiments, ASME 2003 Design engineering technical conferences, Chicago, Illinois, USA, Sept. 2-6.

Jin, Rulchen, Chen, Wei; Sudjianto, Agus, 2002:On sequential sampling for global metamodeling in engineering design, Proceedings of the ASME Design Engineering Technical Conference, v 2, 2002, p 539-548.

Johnson, M. E., Moore, L. M. and Ylvisaker, D., "Minimax and Maximin Distance Designs," Journal of Statistical Planning and Inference, Vol. 26, No. 2, 1990, pp. 131-148.

Koehler, J. R. , Owen, A. B. 1996: Computer experiments. In Ghosh, S. and Rao, C. R., editors, Handbook of Statistics, 13, 261--308. Elsevier Science, New York.

Leary, S., Bhadksr, A., Keane, A. 2003: Optimal orthogonal-array-based Latin hypercubes. Journal of Applied statistics. 30, 585-598.

Liefvendahl, M. and Stocki, R: A study on algorithms for optimization of Latin hypercubes, http://www.ippt.gov.pl/~rstocki/JSPI-olh.pdf.

Li, W., and Wu,C.F.J. (1997): Columnwise-Pairwise algorithms with applications to the constructionof supersaturated designs, Technometrics, 39, 171-179.

Liu, L. and Wakeland, W.: Combining Sizing Optimizer and Metamodel Optimization for Structural Topology Optimization, Optimization Technology Conference, Breckenbridge, CO, May 10-12, 2004.

Liu, L. Wakeland, W. 2004: Using a sizing-optimizer to optimize topology and shape and partial ground structure approach, AIAA-2004-4520, 10th AIAA/ISSMO Multidisciplinary Analysis and Optimization Conference, Albany, NY, August 30 - September 1.

Mastinu, G.R.M., Gobbi, M. 2001: On the optimal design of railway passenger vehicles, Proceedings of the Institution of Mechanical Engineers, Part F: Journal of Rail and Rapid Transit, v 215, n 2, 2001, p 111-124.

Myers, R. H. and Montgomery, D. C., Response Surface Methodology: Process and Product Optimization Using Designed Experiments, John Wiley & Sons, New York, 1995.

McKay, M.D., Beckman R.J. and Conover W.J. 1979: A comparison of three methods for selecting values of input variables in the analysis of output from a computer code. Technometrics. 21, 239-245 (reprinted in 2000: Technometrics, 42, 55-61.

Morris, M. D. and Mitchell, T. J., 1995, "Exploratory Designs for Computer Experiments," Journal of Statistical Planning and Inference, 43, 381-402.

Nielsen, H.B., Lophaven, S.N., and Søndergaard, J.: DACE: A MATLAB Kriging Toolbox, http://www.imm.dtu.dk/~hbn/dace/

Ohsaki, M., and Swan, C. 2002: Topology and geometry optimization of trusses and frames. Recent advances in optimal structural design. Reston: ASCE.

Osio, I. G. and Amon, C. H., "An Engineering Design Methodology with Multistage Bayesian Surrogates and Optimal Sampling," Research in Engineering Design, Vol. 8, No. 4, 1996, pp. 189-206.

Owen, A. B., 1992, "Orthogonal Arrays for Computer Experiments, Integration and Visualization," Statistica Sinica, Vol. 2, pp. 439-452.

Owen, A.B. (1994): Controlling correlations in latin hypercube samples, Journal of the American Statistical Association, 89, 1517-1522.

Palmer, K. et al., "Minimum bias Latin hypercube design, " IIE Transactions, 2001, 33, 793-808.

Papadrakakis M., Lagaros, N., Tsompanakis, Y., Plevris, V. 2001: Large Scale Structural Optimization: Computational Methods and Optimization Algorithms, Archives of Computational Methods in Engineering State of the art reviews, 8, 239-301.

Papalambros, P. 2002: The optimization paradigm in engineering design: promises and challenges, Computer Aided Design, 34 939-951.
Park, J.-S., 1994, "Optimal Latin-Hypercube Designs for Computer Experiments," Journal of Statistical Planning and Inference, Vol. 39, No. 1, pp. 95-111.

Querin, O., Yong V., Steven, G., and Xie, Y., Computational efficiency and validation of bi-directional evolutionary structural optimisation, Computer Methods in Applied Mechanics and Engineering, Vol. 189, No. 2, 2000, pp. 559-573.

Renner, G. and Ekart, A. (2003), "Genetic algorithms in computer aided design," Computer aided design, 35, 709-726.

Reynolds, D., McConnachie, J., Bettess, P., Christie, W.C., and Bull, J.W., Reverse adaptivity - a new evolutionary tool for structural optimization, International Journal for Numerical Methods in Engineering, Vol. 45, No. 5, Jun, 1999, pp. 529-552.

Rossow H.P. and Taylor J.E., A finite element method for the optimal design of variable thickness sheets, AIAA Journal, Vol. 11, 1973, pp. 1566-1569, in Ref. 5.

Rozvany, G.I.N. and Zhou, M. 1996, Advances in overcoming computational pitfalls in topology optimization, AIAA-96-4113-CP.

Rozvany G.I.N. 20011: Aims, scope, methods, history, and unified terminology of computer-aided topology optimization in structural mechanics, Structural and Multidisciplinary Optimization, 21, 90-108.
Rozvany G.I.N. 20012, Stress ratio and compliance based methods in topology optimization - A critical review, Structural and Multidisciplinary Optimization, 21, 109-119.

Sacks, J., Welch, W.J., Mitchell, T.J. and Wynn, H.P. (1989). "Design and Analysis of Computer Experiments." Statistical Science 4(4): 409-435.

Sacks, J., Schiller, S.B. and Welch, W.J., 1989, "Designs for Computer Experiments," Technometrics, Vol. 31(1), pp. 41-47.

Santner, T., Williams, B, Notz, W. 2003: Design and Analysis of Computer Experiments, 150, Springer-Verlag, New York.

Schramm, U., Thomas, H., Zhou, M., and Voth, B., Topology optimization with Altair OptiStruct, Optimization in Industry II-1999, Vol. 2.

Shewry, M.C. and Wynn, H.P. (1987): Maximum entropy sampling, Journal of Applied Statistics, 14, 165-170.

Simpson, T. W., Mauery, T. M., Korte, J. J. and Mistree, F., "Kriging Metamodels for Global Approximation in Simulation-Based Multidisciplinary Design Optimization," AIAA Journal, Vol. 39, No. 12, 2001, pp. 2233-2241.

Simpson, T. W., Booker, A. J., Ghosh, D., Giunta, A. A., Koch, P. N., and Yang, R.-J. "Approximation Methods in Multidisciplinary Analysis and Optimization: A Panel Discussion," Structural and Multidisciplinary Optimization,

Simpson, T. W., Peplinski, J., Koch, P. N. and Allen, J. K. (2001) "Metamodels for Computer-Based Engineering Design: Survey and Recommendations," Engineering with Computers, 17:2 (129-150)

Simpson, T. W., Lin, D. K. J. and Chen, W. (2001) "Sampling Strategies for Computer Experiments: Design and Analysis," International Journal of Reliability and Applications, 2:3 (209-240).

Sobieszczanski-Sobieski, J. and Haftka, R. T., 1997: Multidisciplinary Aerospace Design Optimization: Survey of Recent Developments, Structural Optimization, 14, 1-23.

Tang, B., 1993, "Orthogonal Array-Based Latin Hypercubes," Journal of the American Statistical Association, Vol. 88, No. 424, pp. 1392-1397.

Tang, B., 1994, "A Theorem for Selecting OA-Based Latin Hypercubes Using a Distance Criterion," Communications in Statistics, Theory and Methods, Vol. 23, No. 7, pp. 2047-2058.

Turner, Cameron J., Campbell, Matthew I., Crawford, Richard H. 2003: Generic sequential sampling for metamodel approximations, Proceedings of the ASME Design Engineering Technical Conference, v 1 A, 2003, p 555-564.

Vanderplaats, G.N. 1999: Structural design optimization status and direction, Journal of Aircraft, 36, 11-20.

Wang, X., Kennedy, D., Williams, F.W,1997: A two-level decomposition method for shape optimization of structures, International Journal for Numerical Method in Engineering, 40, 5-88.

Wang, L. and Thierauf, G., A generalized stress criterion for optimum design, International Journal for Product & Process Improvement, Vol. 1, No. 2, 1999, pp. 154-181.

Welch, W. J., Buck, R. J., Sacks, J., Wynn, H. P., Mitchell, T. J. and Morris, M. D., 1992, "Screening, Predicting, and Computer Experiments," Technometrics, Vol. 34, No. 1, pp. 15-25.

Welch, W. J., Yu, T.-K., Kang, S. M. and Sacks, J., "Computer Experiments for Quality Control by Parameter Design," Journal of Quality Technology, Vol. 22, No. 1, 1990, pp. 15-22.

Vanderplaats, G.N. 1999: Structural design optimization status and direction, Journal of Aircraft, 36, 11-20.

Xie, Y.M. Yang, X., Liang, Q. Steven, G., and Querin, O. 2002: Evolutionary Structural optimization, Recent advances in optimal structural design. Reston: ASCE.

Zhou, M., Rozvany, G.I.N., On the validity of ESO type methods in topology optimization, Structural and Multidisciplinary Optimization, Vol. 21, No.2-3, 2001, pp. 80-83.

APPENIDX

Appendix I

Sample size effect, sampling uniformity, and comparison of 18 experimental designs for computer experiments (2004, attached to the original dissertation)

Appendix II

Sample size, uniformity, and design method comparison for computer experiments (2005, newly added for this publication)

Appendix I

Sample size effect, sampling uniformity, and comparison of 18
experimental designs for computer experiments
(2004, attached to the original dissertation)

Sample Size Effect, Sampling Uniformity, and Comparison of 18 Experimental Designs for Computer Experiments

CONTENTS FOR APPENDIX

ABSTRACT

Could enough samples be more important than better designs for computer experiments? Is more necessarily better than less uniformly distributed sampling for computer experiments? Which design(s) based on different criteria to optimize Latin hypercube design (LHD) may be relatively better for prediction accuracy? This study tried to answer the third question, but more questions including the first two emerged. To optimize the criteria, two global optimization approaches are proposed and employed. The main research methods are simulation tests and statistical analyses based on one-way and two-way ANOVA. The comparison is based on evaluation of the prediction accuracy resulted from different experimental designs generated by minimizing fifteen criteria. Together with these designs, random sampling, Latin hypercube designs, and uniform designs based on the tables on the web are also compared. Kriging models are used for approximating twenty test functions.

The results show that in some cases there are statistically significant differences among the means of the errors resulted from different designs, but more often, the differences are not significant. When there are significant differences, the following are observed. LHD is mostly better than random sampling. Maximin LHD (LHMm) is better than other designs in most of the 2D cases when the sample size is large (18-22). LHMm is almost always worse than LHD and Minimized centered L_2 discrepancy LHD (LHCL2) in 5D cases, and always worse than LHD in 10D cases. LHCL2 is mostly better than LHD and often better than LHMm when the sample size is about 14 or smaller, in 2D cases. The results also show that sampling sizes often have stronger impact on the accuracy than the design types. Sampling size determination should receive much more attention for computer experiments, which may be invaluable to the users. The results also indicate that more uniformly distributed sampling does not necessarily lead to more accurate models. Because of limited resources for the study, the test range is limited. Much more study is needed.

NOTATION

Symbols

d_{ij} the Euclidean distance between points i and j

n (N):   number of design variables

m (M):   number of runs or sample size

q: parameter to be determined

CL_2 centered L_2 discrepancy

ML_2 modified L_2 discrepancy

R   correlation matrix

SL_2 symmetric L_2 discrepancy

$\mathbf{X}_{ij}$ the jth component of the ith sampled point

$\mathbf{U}_{ij}$ the jth element of an independent U [0, 1] (uniform distribution between 0 and 1) random variables, independent of the π_{ij}

y(**x**)   the unknown function of interest

f(**x**)   a known polynomial function of **x**, a global regression model, often a constant term

Z(**x**)   the correlation model

θ_k unknown correlation parameters

$\Phi_{q:}$ design criterion

$\boldsymbol{\pi}_{ij}$ the jth element of ith independent uniform random permutations of the integers 1 through n

117

Design types
1) RandSamp (Rd)   Random sampling
2) LHD   Latin hypercube design
3) LHMm   Maximin Latin hypercube design
4) LHCL2   Minimized CL2 Latin hypercube design
5) LHML2   Minimized ML2 Latin hypercube design
6) LHSL2   Minimized SL2 Latin hypercube design
7) LHOrth   Column orthogonal Latin hypercube design
8) LHMmCL2   Maximin CL2 LHD
9) LHMmML2   Maximin ML2 LHD
10) LHMmSL2   Maximin SL2 LHD
11) LHMmOrth   Maximin column orthogonal LHD
12) LHCL2Orth   CL2 column orthogonal LHD
13) LHML2Orth   ML2 column orthogonal LHD
14) LHSL2Orth   SL2 column orthogonal LHD
15) LHMmCL2Orth   Maximin CL2 column orthogonal LHD
16) LHMmML2Orth   Maximin ML2 column orthogonal LHD
17) LHMmSL2Orth   Maximin SL2 column orthogonal LHD
18) UD   Uniform Design

Test functions
1) AC   Ackley's path function
2) AX   Axis parallel hyper-ellipsoid function
3) DE   De Jong's function 1
4) RB   Rosenbrock's valley function
5) RY   Rotated hyper-ellipsoid function
6) MI   Michalewicz's function
7) BR   Branins's rcos function
8) GD   Goldstein-Price's function
9) SX   Six-hump camel back function
10) PK   Peaks function
11) FR   Froth function
12) HX2 Helix2 function
13) RS   Rose function
14) SG2 Sing2 function
15) SG3 Sing3 function
16) SG4 Sing4 function
17) WD1 Wood1 function
18) WD3 Wood3 function
19) WD5 Wood5 function
20) WD6 Wood6 function

A.1. Introduction

Latin hypercube designs (LHD) have been widely used for computer experiments in many fields. Their projections onto one-dimensional subspaces are evenly distributed and they are relatively easy to generate (Santner, et al, 2003). But they do not always generate desired space-filling designs.

Quite a few criteria or designs have been proposed and employed for generating space-filling designs. To name a few, Shewry and Wynn (1987) introduce maximum entropy sampling. Johnson et al (1990) develop minimax and maximin designs. Owen (1992,1994) and Tang (1992, 1993) consider using orthogonal arrays. Fang and Wang (1994) introduce uniform designs.

Many of the criteria produce experimental designs with some attractive properties, but none of them alone can be completely satisfactory. It would be appealing if one approach could offer all of the desired features of several or all these criteria. (Santner, et al, 2003). One way to try this is to select the design that is the best for the criterion or combined criterion within LHDs.

Several approaches have been proposed to optimize LHD. Morris and Mitchell (1995) extend maximin distance criterion and introduce a Φ_q criterion. Fang et al (2002) optimize LHD by minimizing Centered discrepancy CL2. Iman and Conover (1982) and Owen (1994) minimize a linear correlation criterion for pairwise factors. Tang (1993) selects LHD that is orthogonal after some transformation. There are some other approaches, see e.g. (Santner, et al, 2003; Jin, et al, 2003).

While comparing different approaches for experimental designs or optimizing LHDs, many researchers compare the designs based on the criterion values rather than the goals for computer experiments, e.g. prediction accuracy. This may be problematic. For instance, some people use maximin criterion for evaluating designs. Yet, as is well known and also from the author's simulation tests, maximin designs tend to place points near the boundary of the sampling region (e.g., Santner, et al, 2003), thus the designs are not necessarily space filling designs, not to mention if it serves well the purposes of computer experiments .

A.1.1   Main research questions to be investigated and the purpose of this study

How do different designs based on the criteria serve one of the major goals for computer experiments—prediction or how is the accuracy compared? This study investigates this question with simulation tests on twenty test functions and statistical analyses.

Could it be better to use other criteria than those mentioned above such as Φ_q criterion, CL2, etc. to optimize LHD, e.g. minimizing Modified L_2 discrepancy ML2, Symmetric L_2 discrepancy SL2, column-wise correlation for column orthogonal design, etc., or combinations of these criteria? This study compares these different combining approaches by comparing the errors they generate when they are used for sampling. The metamodels used are Kriging models. They are more powerful and flexible than many other models, e.g. low order polynomials.

The purpose of this study is, through simulation testing and statistical analysis, to see what design approaches constantly produce the lowest prediction error, to see the impact of sample size on the prediction errors, and to see the impact of sampling uniformity on the prediction errors.

A.1.2. Global optimization approaches for optimizing LHD

To optimize the criteria, several approaches have been suggested. Morris and Mitchell (1995) use a simulated annealing (SA). Fang, et al (2002) employ a variant of SA, the threshold accepting algorithm (TA). Park (1994) introduces a row-wise element exchange algorithm. Li and Wu developed a columnwise pairwise program. Jin, et al (2003) proposes an enhanced stochastic evolutionary algorithm. The computational cost of some of these approaches can be very high for large size designs (high dimensions and many runs).

For global optimization, genetic algorithms (GA) have been successfully applied in many fields. In many cases, GA has higher efficiency than SA. So GA is proposed by the author as one of the optimizers to optimize the criteria. Liefvendahl and Stocki proposed to use GA to optimize LHD. The approach is quite different from "standard" GA. The details will be given in the global optimizer section.

Since GA and some other methodologies for optimizing LHD may involve large amount of calculation and not small memory space, an extremely simple approach that does a little calculation and needs tiny memory space might outperform more sophisticated approaches for the purpose. One such an approach will be presented below as another optimizer, which is the stochastic search (SS) or random search.

A.1.3. Layout of this report

At first, the design criteria under investigation are described, followed by a very brief introduction of Kriging model and the twenty test functions used for this study. The next section introduces the two optimization approaches SS and GA. The effect of the parameters in the criteria, q in Φ_q criterion for LHMm design, and the weighting parameters in combining two or three criteria in one approach, are shown.

For comparison of the design types, four types are compared first with ANOVA, which are random sampling, LHD, LHMm, and LHCL2. Then all eighteen types of designs are compared with ANOVA. Last discussed are the impact of sample sizes and sampling uniformity on prediction accuracy.

A.2. Eighteen design types to be compared, Kriging models, and test functions

A.2.1. Eighteen design types including the criteria for optimizing LHD (minimizing these criteria is sought)

1)   Design type 1: Random sampling (RandSamp)
2)   Design type 2: Latin hypercube design (LHD)

The jth component of the ith sampled point is

$$X_{ij} = \frac{\pi_{ij} - U_{ij}}{m}$$

where the π_{ij} is the jth element of the ith independent uniform random permutations of the integers 1 through m (m is the number of samples), and the U_{ij} is the jth element of the ith independent U [0, 1] (uniform distribution between 0 and 1) random variables independent of the π_j (Santner, et al, 2003).

3) Design type 3: Maximin Latin hypercube design (LHMm)

Criterion: Φ_q criterion (slightly modified in the form from that of Morris and Mitchell, 1995)

$$\Phi_q = \left[\sum_{i=1}^{m} \sum_{j=i+1}^{n} d_{ij}^{-q} \right]^{1/q}$$

       n: number of design variables
    m: the number of sampled points and
    d_{ij}:   the Euclidean distance between points i and j
    q: parameter to be determined

4)   Design type 4: Minimized CL2 Latin hypercube design (LHCL2)

Criterion: Centered L_2 discrepancy CL_2 (Hickernell, 1998)

$$[CL_2(P_m)]^2 = (\tfrac{13}{12})^2 - \frac{2}{n} \sum_{k=1}^{m} \prod_{j=1}^{n} [1 + \tfrac{1}{2} | x_{kj} - 0.5 | - \tfrac{1}{2} | x_{kj} - 0.5 |^2]$$

$$+ \frac{1}{n^2} \sum_{k=1}^{m} \sum_{j=1}^{m} \prod_{i=1}^{n} [1 + \tfrac{1}{2} | x_{ki} - 0.5 | + \tfrac{1}{2} | x_{ji} - 0.5 | - \tfrac{1}{2} | x_{ki} - x_{ji} |]$$

    m: number of samples
    n: number of dimensions
    P_m: a set of m points

5)   Design type 5: Minimized ML2 Latin hypercube design (LHML2)

Criterion: Modified L_P discrepancy ML_2 (Hickernell, 1998)

$$[ML_2(P_m)]^2 = (\tfrac{4}{3})^n - \frac{2^{1-n}}{m} \sum_{k=1}^{m} \prod_{l=1}^{n} (3 - x_{kl}^2) + \frac{1}{n^2} \sum_{k=1}^{m} \sum_{j=1}^{m} \prod_{i=1}^{n} [2 - \max(x_{ki}, x_{ji})]$$

    m: number of samples
    n: number of dimensions

6)   Design type 6: Minimized SL2 Latin hypercube design (LHSL2)

Criterion: Symmetric L_P discrepancy SL_2 (Hickernell, 1998)

$$[SL_2(P_m)]^2 = \left(\frac{4}{3}\right)^n - \frac{2}{m}\sum_{k=1}^{m}\prod_{j=1}^{n}(1+2x_{kj}-2x_{kj}^2) + \frac{2^n}{m^2}\sum_{k=1}^{m}\sum_{j=1}^{m}\prod_{i=1}^{n}[1-|x_{kj}-x_{ji}|]$$

m: the number of sampled points and

d_{ij}:   the Euclidean distance between points i and j

P_m: a set of m points

7)   Design type 7: Column orthogonal Latin hypercube design (LHOrth)

Criterion: Orth – maximum value of the correlation coefficients between two different columns in the design matrix

8)   Design type 8: Maximin CL2 LHD (LHMmCL2)

Criterion: $\Phi_q + R * CL2$

R: weighting parameter to be decided

9)   Design type 9: Maximin ML2 LHD (LHMmML2)

Criterion: $\Phi_q + R * ML2$

R: weighting parameter to be decided

10)   Design type 10: Maximin SL2 LHD (LHMmSL2)

Criterion: $\Phi_q + R * SL2$

R: weighting parameter to be decided

11)   Design type 11: Maximin column orthogonal LHD (LHMmOrth)

Criterion: $\Phi_q + R * Orth$

R: weighting parameter to be decided

12).   Design type 12: CL2 column orthogonal LHD (LHCL2Orth)

Criterion: $CL2 + R * Orth$

R: weighting parameter to be decided

13)   Design type 13: ML2 column orthogonal LHD (LHML2Orth)

Criterion: $ML2 + R * Orth$

R: weighting parameter to be decided

14)   Design type 14: SL2 column orthogonal LHD (LHSL2Orth)

Criterion: $SL2 + R * Orth$

R: weighting parameter to be decided

15).   Design type 15: Maximin CL2 column orthogonal LHD (LHMmCL2Orth)

Criterion: $\Phi_q + R * CL2 + R1 * Orth$

R, R1: weighting parameters to be decided

16)   Design type 16: Maximin ML2 column orthogonal LHD (LHMmML2Orth)

Criterion: $\Phi_q + R * ML2 + R1 * Orth$

R, R1: weighting parameters to be decided

17).   Design type 17: Maximin SL2 column orthogonal LHD (LHMmSL2Orth)

Criterion: $\Phi_q + R * SL2 + R1 * Orth$

R, R1: weighting parameters to be decided

18)   Design type 18: uniform designs (UD)

A.2.2 Kriging models

Kriging model is a very powerful and extremely flexible approach for modeling the response surfaces of computer experiments, see e.g. Barton, Simpson, et al. Sacks, et. al (1989) and Welch, et. al. (1992) suggest modeling responses as a combination of a polynomial model plus departure:

$$y(x) = f(x) + Z(x)$$

where y(x) is the unknown function of interest, f(x) is a known polynomial function of x, and Z(x) is the realization of a normally distributed Gaussian random process (one of the options) with mean zero, variance σ^2, and non-zero covariance (other forms can be used also). The f(x) term in the equation is similar to the polynomial model in a response surface and provides a "global" model of the design space; in many cases f(x) is simply taken to be a constant term (Sacks, et al., 1992; Welch et al., 1989, Welch, et al., 1992). But sometimes, the first order or second order polynomials were observed to be better than a constant.

While f(x) "globally" approximates the design space, Z(x) creates "localized" deviations so that the Kriging model interpolates the m sampled data points. The covariance matrix of Z(x) is given by:

$$Cov[Z(x^i), Z(x^j)] = \sigma^2 \mathbf{R}([R(x^i, x^j)]$$

where $\mathbf{R}$ is the correlation matrix, and $R(x^i, x^j)$ is the correlation function between any two of the m sampled data points x^i and x^j. R is a (m by m) symmetric matrix with ones along the diagonal. The correlation function $R(x^i, x^j)$ is specified by the user. Sacks, et al. (1989) and Koehler and Owen (1996) discuss several correlation functions that may be used. A Gaussian correlation function of the form:

$$R(x^i, x^j) = \exp[-\sum_{k=1}^{n_s} \theta_k |x_k^i - x_k^j|^2]$$

is the most frequently used where θ_k is the unknown correlation parameter used to fit the model, and the x_{ki} and x_{kj} are the kth components of sampled points x^i and x^j. In some cases, using a single correlation parameter gives sufficiently good results (Osio and Amon, 1996; Sacks, et al., 1989).

In this study, a DACE program for Kriging modeling developed by mathematicians Nielsen et al was used. The regression models include constant, linear, and quadratic polynomials. In most cases, the constant model was used. The correlation models are Gaussian models with different parameters for different dimensions or coordinates.

A.2.3 Test functions

Most of the test functions listed below are popular test functions for testing global optimization methodologies. The tenth function is Peaks function used in MATLAB. Most of them have high nonlinearity or multi-modes. All of the optimal points in the first group have been confirmed by the author with GA.

Group 1

1)	Function 1(AC): Ackley's path function

$$f(x) = -a * \exp(-b * \sqrt{(1/n)(\sum_{i=1}^{n} x_i^2)}) - \exp(\frac{1}{n}\sum_{i=1}^{n}\cos(cx_i)) + a + e$$

a = 20; b = 0.2; c = 2·pi; I =1:n;
$x_i \in [-30,30]$;

2)	Function 2 (AX): Axis parallel hyper-ellipsoid function

$$f(x) = \sum_{i=1}^{n} i x_i^2$$

$x_i \in [-30,30]$;

3)	Function 3 (DE): De Jong's function 1

$$f(x) = \sum_{i=1}^{n} x_i^2$$

$x_i \in [-30,30]$;

4)	Function 4 (RB): Rosenbrock's valley (De Jong's function 2)

$$f(x) = \left[\sum_{i=1}^{n}100(x_{i+1} - x_i^2)\right]^2 + (1 - x_i)^2$$

$x_i \in [-30,30]$;

5)	Function 5 (RY): rotated hyper-ellipsoid function

$$f(x) = \sum_{i=1}^{n}\sum_{j=1}^{i} x_{ij}^2$$

$x_i \in [-30,30]$;

6)	Function 6 (MI): Michalewicz's function

$$f(x) = -\left(\sum_{i=1}^{n}\sin x_i\right) * \left[\frac{\sin(i x_i^2)}{\pi}\right]^{2m}$$

m=10;
$0 \le x(i) \le pi\ (\pi)$.

7)	Function 7 (BR): Branins's rcos function
$$f(x_1, x_2) = a(x_2 - b x_1^2 + c x_1 - d)^2 + e(1 - f)\cos x_1 + e$$
a=1, b=5.1/(4·pi^2), c=5/pi, d=6, e=10, f=1/(8·pi);
$-5 \le x_1 \le 10, 0 \le x_2 \le 15$.

8)	Function 8 (GD): Goldstein-Price's function
$$f(x_1, x_2) = [1 + (x_1 + x_2 + 1)^2(19 - 14x_1 + 3x_1^2 - 14x_2 + 6x_1x_2 + 3x_2^2)] *$$
$$[30 + (2x_1 - 3x_2)^2(18 - 32x_1 + 12x_1^2 + 48x_2 - 36x_1x_2 + 27x_2^2)]$$
$-2 \le x(i) \le 2$, i=1:2.

9)	Function 9 (SX): Six-hump camel back function
$$f(x_1, x_2) = (4 - 2.1x_1^2 + \frac{x_1^4}{3})x_1^2 + x_1x_2 + (-4 + 4x_2^2)x_2^2$$
$-3 \le x_1 \le 3, -2 \le x_2 \le 2$.

10) Function 10 (PK): Peaks function

$$f(x,y) = 3(1-x)^2 \exp(-x^2 - (y+1)^2) - 10(\frac{x}{5} - x^3 - y^5)\exp(-x^2 - y^2) -$$

$$\frac{1}{3}\exp(-(x+1)^2 - y^2)$$

$$-3 \leq x_1 \leq 3, -3 \leq x_2 \leq 3.$$

Group 2
1)   Froth function
      f= -13+x+((5-y)y-2)y-29+x+((y+1)y-14)y;

2)   helix2 function
      $$f = 10(\sqrt{x^2 + y^2} - 1)$$

3)   rose function
      f= 10(y-x^2)

4)   sing2 function
      f= $\sqrt{5}$ (x-y);

5)   sing3 function
      f= (x-2y)^2;

6)   sing4 function
      f= $\sqrt{10}$ ((x-y)^2) ;

7)   wood1 function
      f= 10*(y-x^2);

8)   wood3 function
      f= $\sqrt{90}$ (y-x^2);

9)   wood5 function
      f= $\sqrt{10}$ (x+y-2);

10)   wood6 function
      f= (x-y)/$\sqrt{10}$

Group 3 (From Floudas, et al, 1999)
1) 2D
a)   Adjiman function (2D) (AD2)
      $$f(x_1, x_2) = \cos x_1 \sin x_2 - \frac{x_1}{x_2^2 + 1}$$
      -1 ≤ x$_1$ ≤ 2, -1 ≤x$_2$ ≤1.

b)   Dixon and Szego function (2D) (DS2)

125

$$f(x_1, x_2) = 4x^2 - 2.1x^4 + \frac{1}{3}x^6 + xy - 4y^2 + 4y^4$$

c)   Dixon and Szego function 2 (2D) (DS22)

$$f(x_x, x_2) = 12x^2 - 6.3x^4 + x^6 - 6xy + 6y^2$$

d)   Rijckaert and Martens function 2 (2D) (RM2)

$$f(x_1, x_2) = 0.5x_1 x_2^{-1} - x_1 - 5x_2^{-1}$$
$$1 \le x_1, x_2, x_3, \le 100$$

2)   3D

a)   Rijckaert and Martens function 3 (3D) (RM3)

$$f(x_1, x_2, x_3) = 0.01x_2 x_3^{-1} + 0.01x_1 + 0.0005x_1 x_3$$
$$1 \le x_1, x_2, x_3, \le 100$$

b)   Rijckaert and Martens function 4 (3D) (RM32)

$$f(x_1, x_2, x_3) = -x_1 + 0.4x_1^{0.67} x_3^{-0.67}$$
$$0.01 \le x_1, x_2, x_3, \le 15$$

3)   4D

a)   Dembo function (4D) (DB4)

$$f(x_1, x_2, x_3, x_4) = 0.4x_1^{0.67} x_3^{-0.67} + 0.4x_2^{0.67} x_4^{-1.67} + 10.0 - x_1 - x_2$$
$$1 \le x_1, x_2, x_3, x_4 \le 10$$

b)   Hock and Schittkowski function (4D) (HS4)

$$f(x_x, x_2, x_3, x_4) = x_1 x_2 - x_1 x_4 - 1250x_2 + 1250x_3$$
$$1000 \le x_1 \le 10000, \ 10 \le x_2 \le 1000, \ 10 \le x_3 \le 1000, \ 10 \le x_4 \le 1000$$

4)   5D

Murtagh and Saunders function (5D) (MS5)

$$f(x_1, x_2, x_3, x_4, x_5) = (x_1 - 1)^2 + (x_1 - x_2)^2 + (x_2 - x_3)^3 + (x_3 - x_4)^4 + (x_4 - x_5)^4$$
$$-5 \le x_i \le 5$$

5)   6D

Alkylation function (6D) (Al6)

$$f(x_1, x_2, x_3, x_4, x_5, x_6) = 1.715x_1 + 0.035x_1 x_6 + 4.0565x_3 + 10.000x_2 + 3000 - 0.063x_3 x_5$$

$1500 \leq x_1 \leq 2000$, $10 \leq x_2 \leq 120$, $3000 \leq x_3 \leq 3500$, $85 \leq x_4 \leq 93$, $90 \leq x_5 \leq 95$, $3 \leq x_6 \leq 12$

(Note: The test results on the third group will be presented in other reports or papers.)

A.3. Two approaches for globally optimizing design criteria

A.3.1. Stochastic search (SS) (Random search)

A. 3.1.1. Approach
A very simple search for the minimal criterion value among those of many LHDs is to randomly generate a LHD and to calculate the criterion. If it is lower than the lowest kept so far, it becomes the new lowest; otherwise it is ignored and a new LHD is generated and evaluated. The process is repeated until some stopping criterion is met. This is the stochastic search (SS) or random search.

A variation to SS is to keep tract of LHDs generated. The criterion is only evaluated for the new LHDs. This process increases the memory space and operation for comparing LHDs. Unless this increase of the operation and space is smaller than those for evaluating the criterion, the first approach may be more efficient than the second approach. The first is used in this study.

Comparing with many other optimization methods for optimizing LHDs, SS involves much less calculation and much less memory space, thus SS might be able to outperform them in some cases.

A.3.1.2. Stopping criterion: convergence populations
Talking about random search, many people often think the criterion searched would not be significantly improved or the best could not be found until the final option is checked or all of the options are exhausted. Let us examine the behavior of SS for optimizing the criteria for this study.

The stopping criterion is the population size that means how many random LHDs are generated and evaluated before stop. One-way and two-way ANOVA were used to compare the means of the lowest criterion values generated by employing different population sizes, for each type of the designs. For every criterion, (10, 100, 1000, 2000,..., 10000) were used as population sizes for the first testing. Then, for some criteria, (1000, 2000, ..., 10000) were used for the second testing. Further, (2000, 3000,..., 11000) or (3000,4000,..., 12000), or (4000, 5000,...,13000), were used, to see if there is still any statistically significant difference. A group of typical output for LHMm is shown below. For all other criteria, the output can be provided.

> ANOVA box plots for LHMm (m: number of runs; n: number of design variables)
> 1)   m=10, n=2
>   a) population size: 10, 100, 1000, 2000,....,10000 (corresponding to the column numbers)
>   p-value = 0.0000

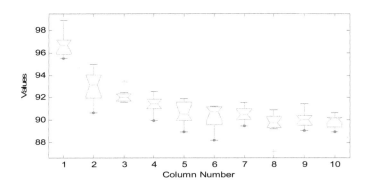

b) population size: 2000, 3000, ...,11000
   p-value = 2.8706e-004

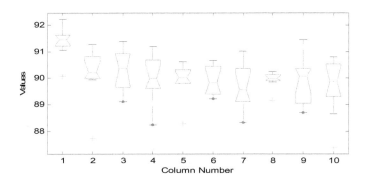

c) population size: 3000, 4000, ...,12000
   p-value = 0.0400

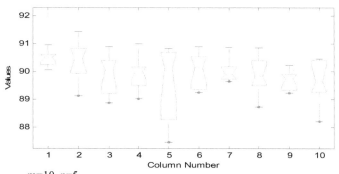

2)   m=10, n=5
    a) population size: 1000, 2000,….,10000
       p-value = 1.2021e-005

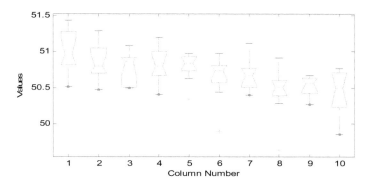

    b) population size: 2000, 3000, ...,7000
       p-value = 0.0155

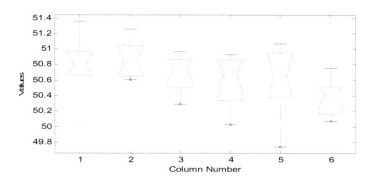

3) m=15, n=10
   population size: 10, 100, 1000, 2000,....,10000
   p-value = 0.0

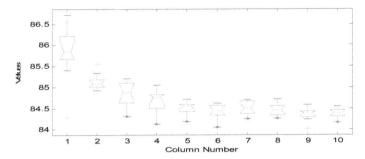

For most criteria, starting from 3000 as the population size, no significant difference in the means is observed, except LHMm for which 4000 is needed for convergence.

It is obvious for each sampling, SS will drive criterion lower and lower with more and more runs. What the output shows above is that the means of the criterion values of many samplings will go down quickly initially. The down speed becomes smaller and smaller, and does not show significant down signs after passing 3000 or 4000 (population sizes) or starts fluctuation within a small range. So it is possible, instead of been exhausted, the search can be stopped after generating three or four thousand samplings, and the criterion found might be very close to the best one.

A.3.2   Genetic algorithms (GA)
         Genetic algorithms (Holland 1975, Goldberg 1989) are popular tools for global optimization and have been widely used in many fields. GA is a stochastic global search method. It operates on a population of potential solutions, applying the principle of survival of the fittest to produce better and better solutions. At each generation, a new population is created by selecting individuals according to their relative fitness and breeding them using crossover and mutation. (Genetic Algorithm Toolbox: User's guide, by Chipperfield, et al).
         The GA program used in this study is based on the Genetic Algorithm Toolbox for general use, developed by Andrew Chipperfield et al, which has gone through extensive testing and application in many fields worldwide. This section cites many statements from the guide. The core operations crossover and mutation were developed or modified by the author (Liu).

A.3.2.1.Representation
         Every population is represented by a matrix, here called population matrix by the author. Each row is an individual representing a Latin hypercube sampling (m points), which is usually a matrix itself, here called LHD matrix. The row in the population matrix is formed by changing the Latin hypercube sampling matrix into a row vector ordered column-wise. That is, (1,1) element in the LHD matrix is (1,1) element in the population matrix; (2,1) element in LHD matrix is (1,2) element in the population matrix; etc.

131

The number of rows in the population matrix is called population size, showing how many individuals (LHDs) are in the population. The number of columns is the number of all the elements in one LHD sampling matrix.

Instead of using binary-code representation, real-valued representation is used in this study to take the advantages mentioned in the guide, e.g. efficiency.

A.3.2.2. Selection

Selection is to select the individuals for reproduction from the population, based on their relative fitness. In this study, a stochastic universal sampling (SUS) is employed, which is a single-phase sampling algorithm with minimum spread and zero bias, see the guide for details. SUS uses N equally spaced pointers, where N is the number of selections required. The population is shuffled randomly and a single random number in the range of [0 Sum/N] is generated, ptr. Sum is a real-valued interval determined either as the sum of the individuals' expected selection probabilities or the sum of the raw fitness values over all the individuals in the current population. The N individuals are then chosen by generating the N pointers spaced by 1, [ptr, ptr+1,...,ptr+N-1], and selecting the individuals whose fitness span the positions of the pointers.

A.3.2.3. Crossover

Ordinary crossover operators cannot be used since they would disturb the Latin hypercube designs. To stay in the domain of LHD, crossover is designed to exchange the elements between two neighboring individuals in the population matrix, which represent the two columns in the two LHD matrix, with the same column number. Since, each column in each LHD is an independent random permutation, the new individuals generated by crossover are still LHDs. One or more pairs of columns are exchanged by crossover in this way, based on the probability for crossover selected by the user. Some other optimization methods for optimizing LHDs, e.g. simulated annealing, do not use exchanging columns, only exchanging elements within one column that is the mutation used by this author as explained below.

A.3.2.4.   Mutation

Mutation is exchanging two elements within one column of the LHD matrix. The column is randomly selected, so are the elements within the selected column. The user selects the probability for mutation.

A.3.2.5.   GA parameters

The users of GA need to select several parameters that might have large impact on effectiveness and efficiency. Among those for this study, four parameters seem important. They are population size (Pop), maximum generation (MaxG) for stopping control, the probability for crossover (Pc), and the probability for mutation (Pm). The User's Guide recommends 30 –100 for population size, uses several hundreds for MaxG, uses the value around 0.7 for Pc, and uses the value about 0.01 for Pm for some examples. All the parameters can and should vary depending on the need for the problem in hand.

For this study, since crossover and mutation are quite different from those in general GA, better Pc, and Pm as well as Pop and MaxG need to be found. To optimize the four parameters, metamodeling optimization was used to optimize them simultaneously.

The problem formulation for optimizing the four parameters follows:

The design variables: Pop, MaxG, Pc, Pm
The objective functions: criterion + R * time

132

Criterion: as listed in the first section, either single or combined, the mean value of a group of the output of ten samples was used;

R: weighting factor.

Time: time used for sampling, the mean of ten samples was used.

Model: Kriging model was used for approximating the output of the objectives;

Optimizer: GA or GA followed by sequential quadratic programming (SQP).

The optimization needs a lot of time. It was planned for 2, 5, and 10 design variables and for every design type. The better values found so far were used in GA operations for the main tasks of this study. The following ranges were used for the parameters. For different design types, different values within the ranges were used.

For Pop: 10 –90;
For MaxG: 100 –500;
For Pc: 0.1 – 0.8;
For Pm: 0.3 – 0.85.

Some preliminary testing has been done to compare SS and GA. It seems GA is more effective in many cases, but sometimes SS can find better solutions within shorter time. Much more testing is needed to get better understanding.

Some typical optimization process plots for optimizing the criteria by GA are shown below.

For optimizing LHML2Orth:

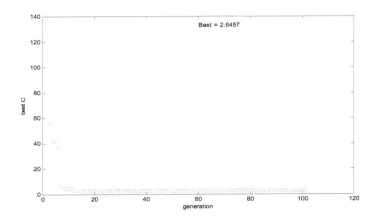

For optimizing LHMmSL2Orth:

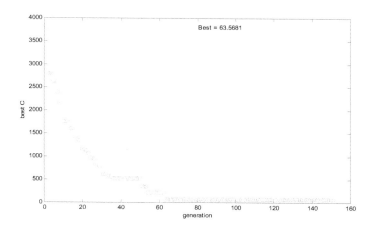

A.4. The effect of the parameters

A.4.1 q effect in Φ_q criterion

For Maximin LHDs (LHMm), the Φ_q criterion has a parameter q. Morris and Mitchell (1995) recommend to try different q values and use the one that results in better Maximin criterion values. As mentioned before, in this study, it is desired to see how different q values affect the prediction accuracy. Prediction is one of the typical purposes for computer experiments.

All of the ten test functions in the first group were used for testing the effect of q. Either GA or SS was used for optimizing the Φ_q criterion. The number of design variables is 2, 5, or 10. The q values always include 1, 2, and 5 as the first three values, all other 10 values were randomly generated between 6 and 100. The test results are shown below.

1)   N=2, m=4,8,12,16,20.

The q-values tested (each row corresponds to one of the ten function in Group 1)

1	2	5	8	21	26	43	50	53	67	73	82	98
1	2	5	7	21	29	38	48	55	66	75	88	99
1	2	5	12	21	33	34	48	58	65	77	86	95
1	2	5	8	18	24	43	50	56	66	78	85	92
1	2	5	6	20	29	37	49	59	62	74	85	96
1	2	5	10	15	24	35	49	56	68	76	87	91
1	2	5	6	22	33	42	46	58	68	76	85	95
1	2	5	12	21	33	34	48	58	65	77	86	95
1	2	5	8	18	24	43	50	56	66	78	85	92
1	2	5	6	20	29	37	49	59	62	74	85	96

The p-values of the one-way ANOVA

0.6789	0.3602	0.0347	0.4534	0.9457
0.0129	0.2556	0.5916	0.4703	0.5075
0.0017	0.0000	0.0175	0.0652	0.4535
0.1516	0.5105	0.9898	0.3611	0.5173
0.0148	0.9491	0.1877	0.7361	0.5911
0.1072	0.7294	0.6086	0.4047	0.0775
0.8957	0.6768	0.4080	0.0907	0.5138
0.0798	0.2258	0.1255	0.3803	0.5071
0.4191	0.0812	0.0559	0.0213	0.0688
0.0048	0.3152	0.0828	0.3973	0.8264

In most cases, there is no significant difference in the means of the RMSE (root mean square error) from using different q values. Two out of five cases for function DE have significant differences. The q values of 5 and 21 appear to result in relatively lower errors in one or two cases. The value of 5 is used for DE 2D comparison studies presented in Section A.5.

2)   N = 5, m = 5, 15, 25, 35, 45.

The q-values tested _(each row corresponds to one of the first five functions in Group 1)

1	2	5	10	18	33	39	48	58	68	73	89	94
1	2	5	10	15	31	36	47	58	62	76	90	92
1	2	5	12	23	33	43	47	59	63	76	86	93
1	2	5	9	16	29	37	51	60	71	77	87	93
1	2	5	12	24	25	39	43	58	65	72	88	97

The p-values of the one-way ANOVA

0.0119	0.0000	0.0001	0.0002	0.0000
0.0000	0.4474	0.0244	0.0146	0.0009
0	0.1278	0.0327	0.0000	0.0000
0.0000	0.0000	0.0000	0.0005	0.0000
0.1324	0.0192	0.5669	0.1705	0.1288

By comparing the RMSE means, it appears that $q = 75$ result in lower errors in many cases for the first three functions, and $q = 45$ for the other two functions, so they are used in comparison tests in Section A.5.

3) $N = 10$, $m = 10, 30, 50, 70, 90$.

The q-values tested _(each row corresponds to one of the first five functions in Group 1)

1	2	5	9	22	27	36	50	53	67	74	82	91
1	2	5	12	19	31	38	51	54	70	75	82	98
1	2	5	8	20	27	38	48	60	68	75	88	99
1	2	5	12	18	27	40	50	57	66	74	83	93
1	2	5	9	23	32	37	45	57	63	81	90	93

The p-values of the one-way ANOVA

0.0000	0.0000	0.0000	0.0000	0.0000
0	0	0.0000	0.1483	0.4593
0	0	0.0067	0.4553	0.0116
0	0.0000	0.0081	0.4842	0.1982
0.5193	0.2103	0.1837	0.3961	0.0225

By comparing the RMSE means, it appears that $q = 75$ result in lower errors in many cases for the first four functions, so it is used in comparison tests in Section A.5. For RY function, there is no case with p-value below 0.01. Checking the RMSE means shows that $q = 1$ results in lower errors in the last case, so $q = 1$ is used for comparison tests in Section A.5. And, $q = 75$ was tested on function RY that resulted in higher error levels than using $q = 1$.

A.4.2.   R, R1 effect for combined criteria
    It is desirable to find the best weighting parameters R, R1 so that the combined criteria could result in better designs. At first, optimization was tried and then it was noticed that these parameters might not have significant effect on the errors. Some preliminary tests were done on some combined criteria. Some results are shown below. Strictly speaking, the

combined criteria result in multi-objective optimization problems (MOO or MDO). The active research and new development in this area could be used to optimize the combined criteria or designs. Pareto front might need be found especially for the objectives that conflict with each other or there is some curve on which one objective cannot be improved without causing the other worse. This topic may belong to future work.

One way and two-way ANOVA were employed for testing the effect of R and R1. Eight R (R1) values were generated by stratified random sampling between 0.0001 and 10000. Sampling was done using the designs, followed by Kriging modeling. Errors were found through verification on 100 points within the domain. The relative root mean square error (RMSE divided by the mean of function values) was used for comparison. Different test functions, different design types, different numbers of design variables, and different run numbers were tested. It seems that in most of the cases tested, there is no significant difference by using different R (R1) values. Occasionally, there is some significant difference. But none of R (R1) value(s) dominates. The following are some output plots of the one-way ANOVA. Pv1 (pv1) is the probability value of the one-way ANOVA. Pv2 (pv2) is the probability value of the two-way ANOVA.

1) Design criterion: LHMmCL2
   Test function: Rosenbrock's valley function
   Number of design variables: n = 2
   Number of runs: m = 4
   Pv1= 0.2299

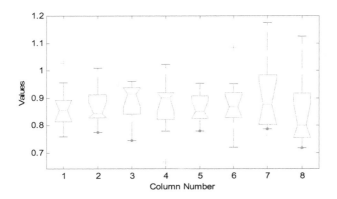

2) Design criterion: LHSL2Orth
   Test function: De Jong's function 1
   Number of design variables: n = 2
   Number of runs: m = 4, 6, …,22

137

R values tested

R=
1.0e+003 *

0.00000062318614
0.00000163084832
0.00004513443870
0.00037870006712
0.00636069029062
0.05031131783713
0.66566014159444
5.78561952901596

pv1 =

0.0018
0.7801
0.2616
0.3896
0.3884
0.8269
0.1552
0.2680
0.5779
0.5397

pv2 =   0.0033   0   0.7338

Pv1 shows only one case (10%) has significant difference. In Pv2, the first data shows column (R, R1) effect, the second row (sample size) effect, and third interaction between columns and rows.
The ANOVA Plot for the first case:

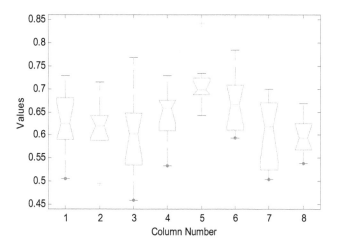

3) Design criterion: LHMmSL2
   Test function: Rotated hyper-ellipsoid function
   Number of design variables: n = 5
   Number of runs: m = 7,9,...,25
          R values tested

   R=
   1.0e+003 *
     0.00000061507023
     0.00000315694510
     0.00008415026296
     0.00042892885528
     0.00157677837473
     0.03278284403785
     0.78471206883702
     9.60323899650889

   pv1 =
     0.7856
     0.0045
     0.3205
     0.0442
     0.4510
     0.4531
     0.5359
     0.4014
     0.4771
     0.6420

pv2 = 0.9041   0   0.081

Pv1 shows that only the second case (10% of all the cases) has significant difference. In Pv2, the first data shows column effect, the second row effect, and third interaction between columns and rows.

The ANOVA Plot for the second case:

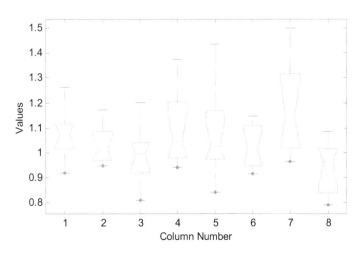

4) Design criterion: LHMmCL2Orth
   Test function: Peaks function
   Number of design variables: n = 2
   Number of runs: m = 4,6,...,22

   R values tested
   R, R1=
   1.0e+003 *
0.0004	0.0093
7.3958	0.0933
0.0032	0.0000
0.0000	0.1098
0.0899	0.0000
0.0000	6.8653
0.0000	0.0007
0.8490	0.0000
0.0011	0.0000
0.0006	6.3746
0.0334	0.0007
1.3000	0.0498
0.0000	0.0081
0.0000	0.4619
0.0001	0.0000
0.3137	0.0000

pv1 =

0.4650
0.0066
0.3312
0.4744
0.3853
0.0194
0.2846
0.4886
0.0053
0.4735

pv2 =0.0162   0.3838   0.4140

Pv1 shows only two cases have significant difference. In Pv2, the first data shows column effect, the second row effect, and third interaction between columns and rows.

The ANOVA Plot for the second case:

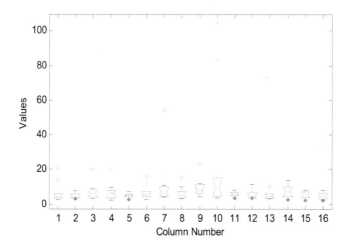

The ANOVA Plot for the ninth case:

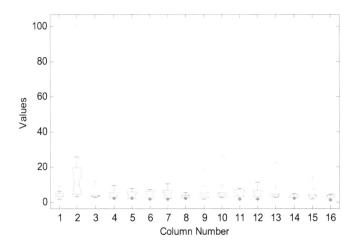

In most cases tested, there is no significant difference in the means of the relative squared errors resulted from different R (R1) values. Thus, in all the following tests, R =1 and R1 = 1 were used. Again, for these multiple criteria, much more investigation is needed to find better combinations of the criteria for better designs.

A.5.   Comparison of 18 design types for prediction accuracy

In this section, the research method is described first and then the results are presented.

The main study tools employed are simulation tests and statistical analysis based on one-way and two-way ANOVA (analysis of variance). An experimental design method is employed for sampling. Kriging models are used to approximate twenty test functions. Sampling for verification is conducted by Latin hypercube design to sample 5000 or 10000 points within the design space. The accurate responses and those generated by Kriging models are compared to find root mean square errors (RMSE), maximum errors, and relative RMSE (RMSE divided by the mean of the responses of all the response for verification). This process is repeated at least ten times up to 20 times for the same pair of dimensions and sample sizes. The above process is repeated for each design type. Next, the sample size is increased and the process is repeated. Finally, the entire process is repeated for another function. The numbers of dimension (n) tested are 2, 5, and 10. The sample size vary from 4 to 100.

One way and two-way ANOVA are employed to compare different designs by comparing the means of RMSE to see if there is any statistically significant difference among the means generated by different designs. Also in some cases, the means of the maximum errors are compared, with similar results to those of RMSE. The critical probability value or p-value used for this study is 0.01, or 1% that is usually a conservative p-value. The results with 5% are also shown.

The above testing and ANOVA process is repeated for each test function separately, so each comparison is for the same test function. There is no mixed comparison between different test functions except that for comparing the outputs from different sampling uniformity degrees. The process for uniformity is as follows. At first, a sampling is done; a Kriging model is built to approximate a test function; verification is conducted to find the errors. With the exactly the same sampling, another Kriging model is built to approximate another test function and to find the errors by comparing with the exact response. This process is repeated twenty times for twenty functions respectively. Then, ANOVA is conducted to compare two different designs by comparing the means of the errors of approximating twenty functions generated from these two designs. The verification points are exactly the same for every function and for every design type.

For one-way ANOVA, the explanation of the box plots follows. The centerline is the median; the bottom and top lines of the box are the quartiles; the end lines of the whiskers are the ends of the data range. The outliners are those marked as "+". The dots on the end lines show that there is no outliner.

A.5.1   Comparison among random sampling, LHD, LHMm, LHCL2 by GA

It is desirable to have comprehensive comparison among all the design types under investigation. But because of the extremely limited resource, only four design types listed in this section underwent quite extensive testing as is shown below.

All the twenty test functions were used for testing 2D cases, the first five functions for 5D and 10D cases. The results from using SS and GA are similar. The result of using GA is reported here. For each n (dimension) and m (run number) values, four designs were used for sampling, the Kriging model was built for each design, and then errors were found through verification by using 10,000 points sampled by LHD.

From the test results, LHD is nearly always better or not worse than random sampling. Maximin LHD (LHMm) is better than other designs in most of the 2D cases when the sample size is large (18-22). LHMm is mostly worse than LHD and Minimized centered L_2

The critical p-value is selected as 0.01 that is a quite conservative but not too conservative critical value after comparing several p-values, F values and ANOVA plots. Two of the one-way ANOVA plots with p-values close to 0.01 and 0.05 are shown below.

p= 0.0102

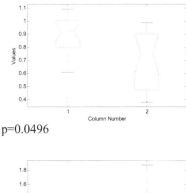

p=0.0496

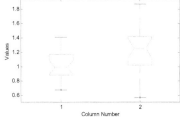

A.5.1.1 n = 2, m = 4, 6, …, 22. (m: number of runs; n: number of design variables)

The number of points for verification is 10,000 generated by LHD.

Among all 200 cases (there are 20 sampling groups in each case), 86 or 43% of the total cases have one-way ANOVA p-values below 0.01, 109 or 54.5% cases below 0.05. In most of these cases, LHD is better than random sampling. LHMm is sometimes better than random sampling and LHD, but worse in the other cases. LHCL2 is mostly better than random sampling and LHD, better than LHMm in more cases with the opposite in fewer cases, when there are significant differences between LHMm and LHCL2.

Since Random sampling and LHD behaves so close, it is desired to test the two only and to get the exact number of cases that LHD is significantly better than random sampling. For each pair of n and m values, 20 designs were sampled, the Kriging model was built for each design, and then errors were found through verification by using 10,000 points sampled by LHD. The verification sites are exactly the same for both design groups. Similarly, pair-wised tests between LHD and LHMm, LHD and LHCL2, and LHMm and LHCL2 were conducted.

Among 200 cases, the p-values for one-way ANOVA in 17 cases or 8.5% are below 0.01. In all the 16 cases, LHD is significantly better than random sampling. When 0.05 is used as the critical value, 48 or 24% cases have p-values below 0.05. LHD is better than RandSamp in 46 cases.

Among 200 cases for comparing LHD and LHMm, 54 cases (27%) have significant differences (p<0.01). In 27 cases, LHD is better; in 27 cases, LHMm is better.

Among 200 cases for comparing LHD and LHCL2, 17 cases (8.5%) have significant differences (p<0.01). In only 2 case, LHD is better; in 15 cases, LHCL2 is better.

Among 200 cases for comparing LHMm and LHCL2, 78 cases (39%) have significant differences (p<0.01). In 30 cases, LHMm is better (mostly happened when the sample size is from 18 to 22); in 48 cases, LHCL2 is better (including the cases when m=18, 20, 22).

Table A.1 summarizes all the results including those for n=5 and 10.

Table A.1. Summary of the tests comparing Random Sampling, LHD, LHMm, LHCL2

Comparing 4 design types Rd, LHD, LHMm, LHCL2					
1)	Compare RD LHD LHMm LHCL2				
			N=2	N=5	N=10
			200 cases	50 cases	50 cases
			Percentage, %	Percentage, %	Percentage, %
	p-v <0.01		43.5	14	26
	p-v <0.05		54	24	42
2)	Compare Rd LHD				
	p-v <0.01		8.5	2	4
		Rd better	0.5	0	0
		LHD better	8	2	4
	p-v <0.05		24	2	14
		Rd better	1	0	2
		LHD better	23	2	12
3)	Compare LHD LHMm				
	p-v <0.01		27	10	28
		LHD better	13.5	10	28
		LHMm better	13.5	0	0
	p-v <0.05		42.5	22	48
		LHD better	25	22	46
		LHMm better	17.5	0	2
4)	Compare LHD LHCL2				
	p-v <0.01		8.5	8	2
		LHD better	1	6	0
		LHCL2 better	7.5	2	2
	p-v <0.05		15.5	20	12
		LHD better	1.5	12	10
		LHCL2 better	14	8	2
5)	Compare LHMm LHCL2				
	p-v <0.01		39	12	30
		LHMm better	15	2	2
		LHCL2 better	24	10	28
	p-v <0.05		51.5	28	46
		LHMm better	21.5	6	4
		LHCL2 better	30	22	42

Explanation:
Design types compared: Rd (random sampling), LHD (Latin hypercube design), LHMm (Maximin LHD), LHCL2   (minimized center discrepancy LHD).
N: dimensions.
Percentage: the number of the cases where the different designs are not equivalent (at the given p-value) divided by the total number of cases in the group.
p-v: probability value from ANOVA (one-way).

The following figures show the changes of RMSE and maximum errors with the sample sizes for each design types and for every function. Each point is the average of 20 values.

146

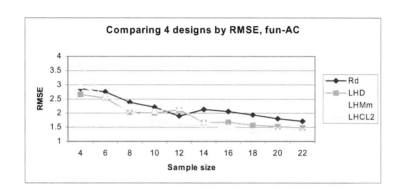

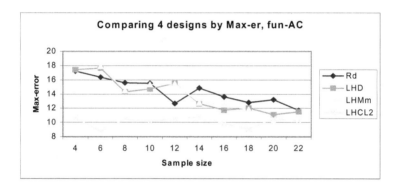

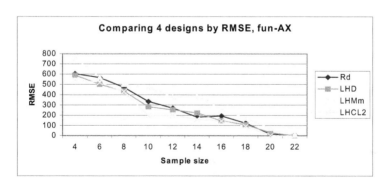

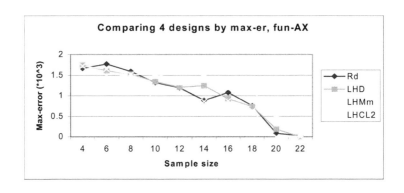

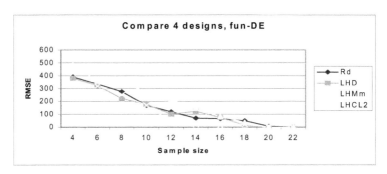

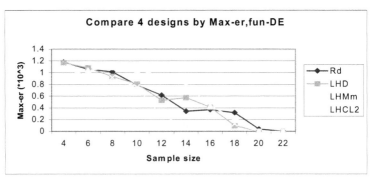

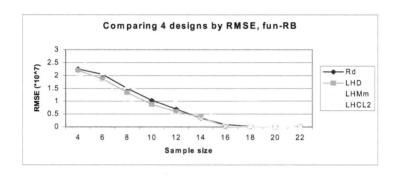

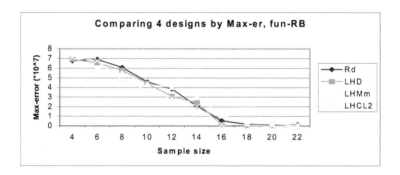

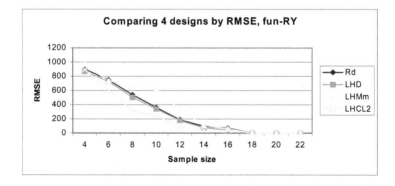

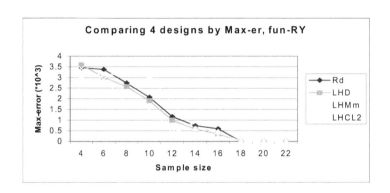

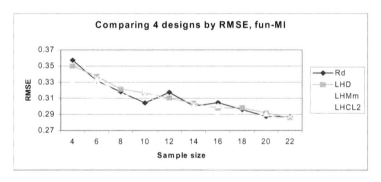

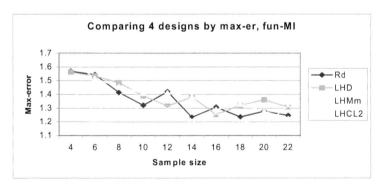

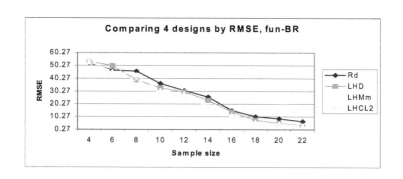

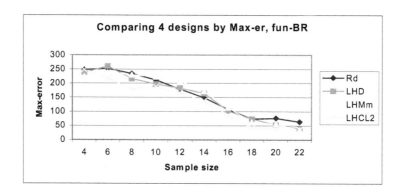

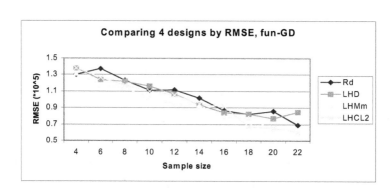

151

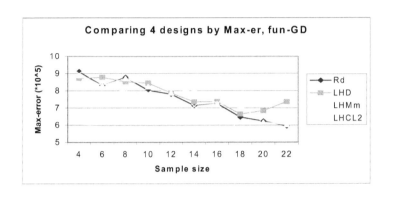

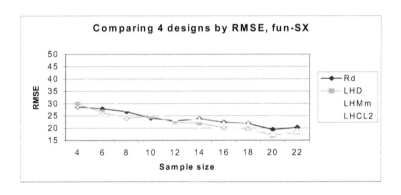

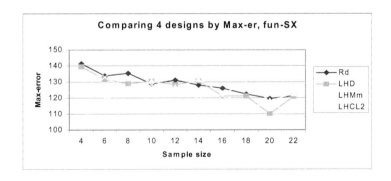

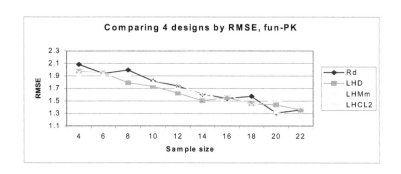

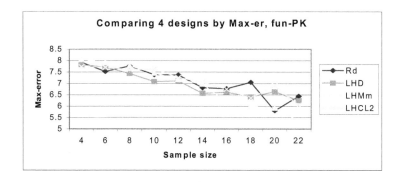

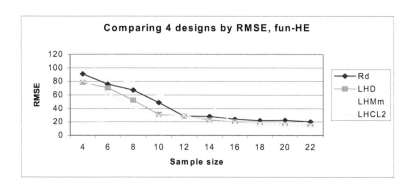

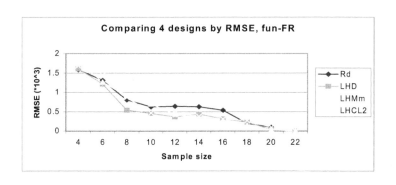

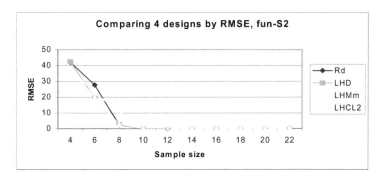

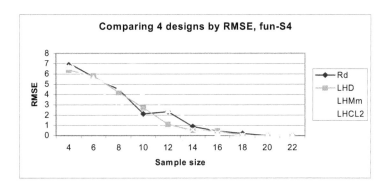

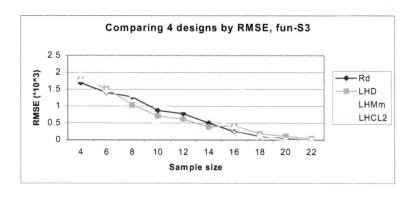

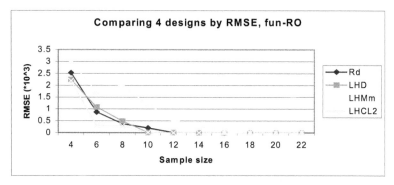

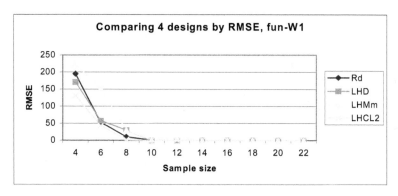

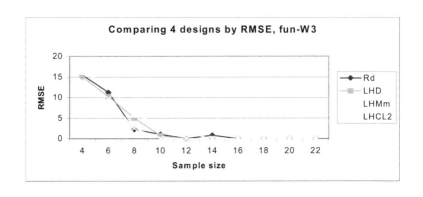

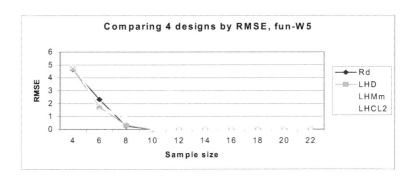

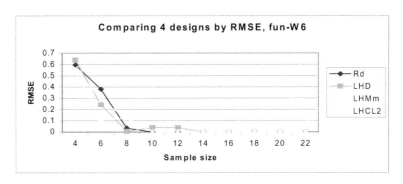

A.5.1.2 n = 5, m = 5,10,15, ...,50 (m: number of runs; n: number of design variables)
The number of points for verification is 5000, generated by LHD.

Among all 50 cases, 7 or 14% of the total cases have p-values below 0.01; 12 or 24% cases have one-way ANOVA p-values below 0.05.

In order to get the precise numbers of the cases with statistically significant differences between the two design types, pair-wised tests were done.

Among 50 cases for comparing random sampling and LHD, 1 cases have significant differences (p-value < 0.01). LHD is better in the case.

Among 50 cases for comparing LHD and LHMm, 5 cases have significant differences (p-value < 0.01). In all these cases, LHD is better.

Among 50 cases for comparing LHD and LHCL2, 4 cases have significant differences (p-value < 0.01). LHD is better than LHCL2 in 3 cases, opposite in the other case.

Among 50 cases for comparing LHMm and LHCL2, 6 cases have significant differences (p-value < 0.01). In one case, LHMm is better; in 5 cases, LHCL2 is much better.

Table A.1 summarizes the result.

The following figures show the changes of RMSE with the sample sizes for each design types and for every function. Each point is the average of 15 values.

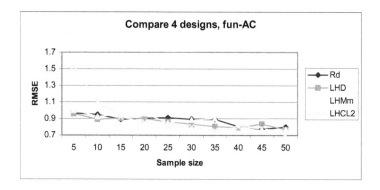

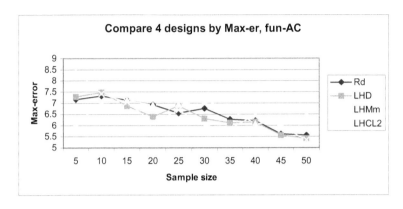

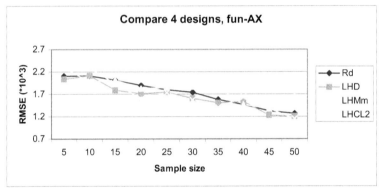

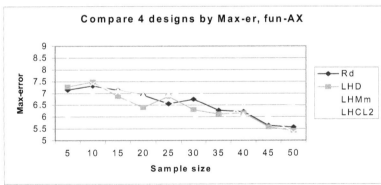

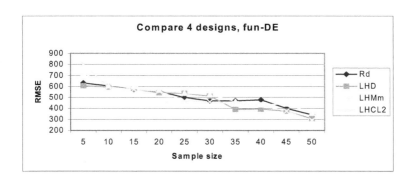

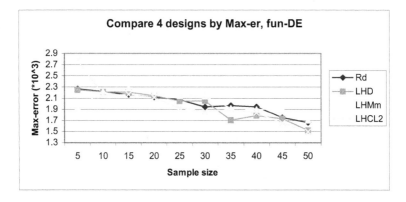

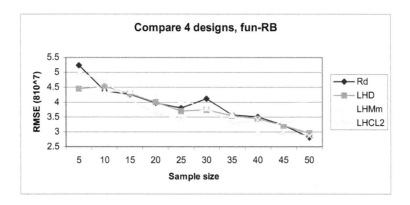

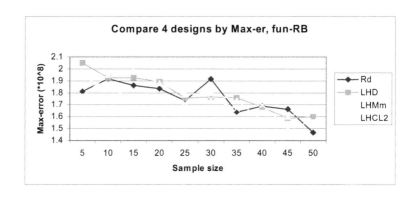

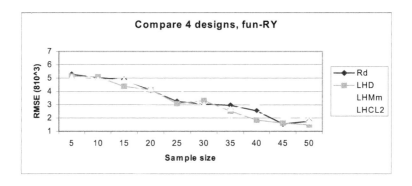

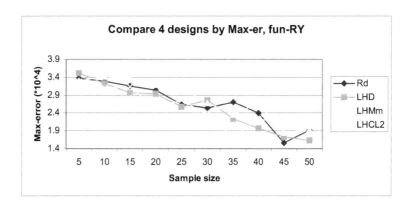

A.5.1.3   n = 10, m = 10, 20, ...,100 (m: number of runs; n: number of design variables)

The number of points for verification is 5000, generated by LHD.

There are 13 cases, or 26% showing p-values below 0.01; 21 cases or 42% show p-values below 0.05. The means of RMSE (means of the 10 groups with the same dimensions and sample size) are shown below.

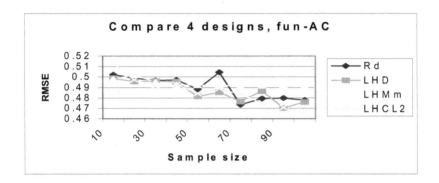

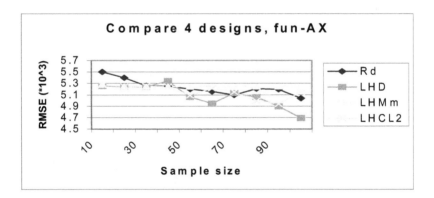

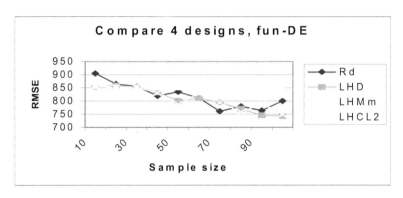

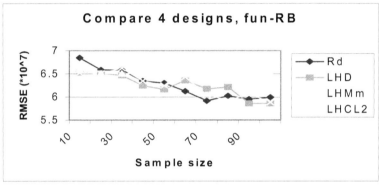

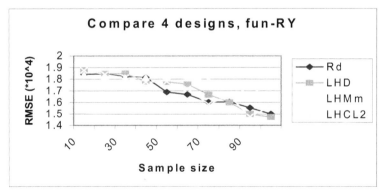

A.5.2.   Comparison among 18 design types
        To evaluate 15 approaches for optimizing Latin hypercube designs together with Random sampling, LHD, and uniform design, 18 design types were compared with one-way and two-way ANOVA. Only the first ten functions were tested.

5.2.1 n = 2, m = 4,6,...,22. (m: number of runs; n: number of design variables)
        The number of points for verification is 5000, generated by LHD. Out of total 100 cases, 37 cases or 37% show significant differences (p-value < 0.01) in the means of the errors generated by different designs; 60 cases show p-values below 0.05. The means of RMSE (means of the 10 groups with the same dimensions and sample size) are shown below.

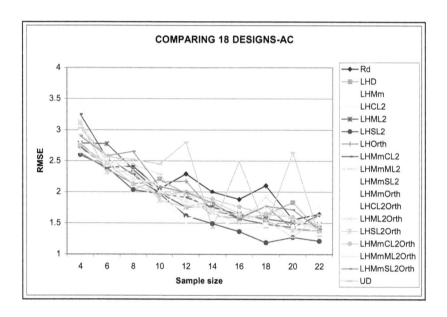

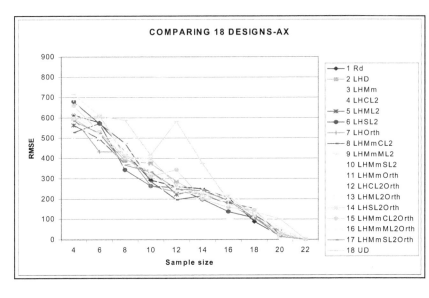

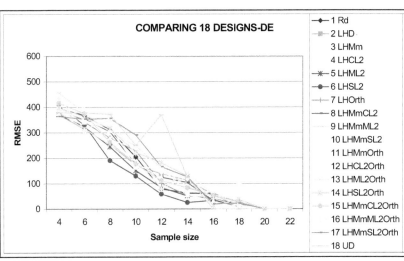

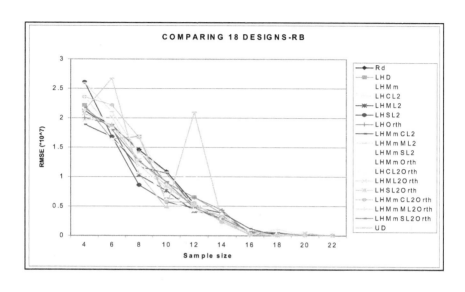

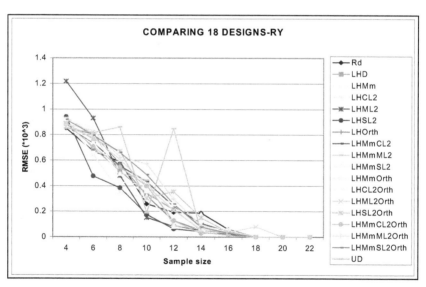

165

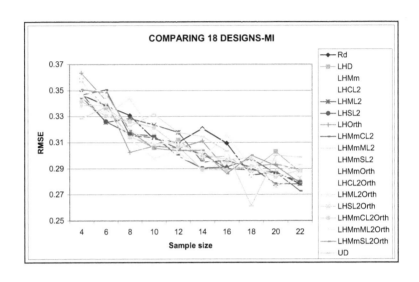

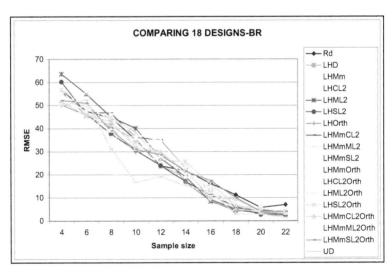

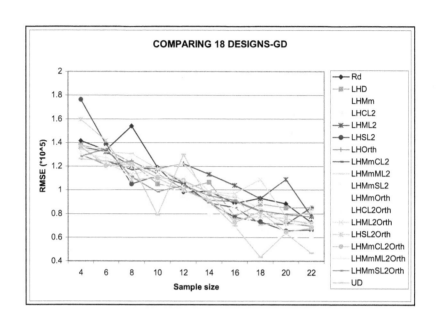

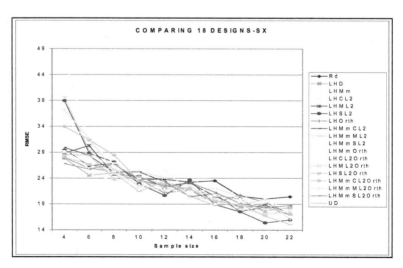

167

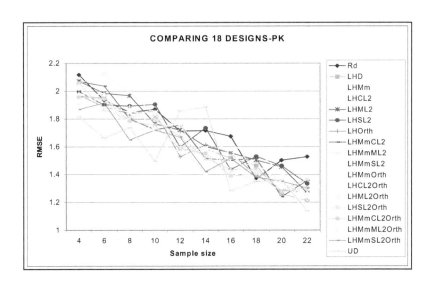

A.5.2.2 n = 5, m = 8,11,...,20 (m: number of runs; n: number of design variables)

The number of points for verification is 100, generated by LHD. In all 50 cases, 8 cases or 16 % have significant differences (p-values < 0.01); 13 cases, or 26% of the total cases have p-values below 0.05.

Table A.2. Summary for tests to compare 18 design types

p-values	N=2		N=5	
	quantity	Ratio, %	quantity	Ratio, %
p-v <0.01	37	37	8	16
p-v <0.05	60	60	13	26
Total case number	100	100	50	100

A.6. Could enough samples be more important than better designs?

In the majority of the cases tested, there is no significant difference (p-value < 0.01) between the error means generated by different design types. Even when there are significant differences in other cases, it seems that adding several runs could achieve better accuracy than that by "better designs" in 2D cases sometimes. The examination of some mean output for the relative root mean square error shows that there are larger differences among the results caused by different run numbers than those caused by different designs. The examination of the two-way ANOVA tables reveal that the portion of the variation in terms of the sum of the squares caused by different sample sizes is often much larger than that caused by different designs. The following matrix (Table A.3), figure, and ANOVA table (Table A.4) shows the case for Branins' rcos function, while comparing four designs random sampling, LHD, LHMm, and LHCL2, with 2 design variables. Each element is the average of 20 output values (the relative RMSE). It is more similar across the columns than across the rows. It shows that run number has larger impact on the error means. More similar figures are shown in the last section for 2D, 5D, and 10D cases.

Table A.3 Average RMSE values for comparing Rd, LHD, LHMm, and LHCL2

m	RD	LHD	LHMm	LHCL2
4	1.0125	0.9795	0.9833	1.1204
6	0.8597	0.9356	0.8427	0.8615
8	0.7604	0.7256	0.7576	0.6805
10	0.7009	0.5654	0.6325	0.5974
12	0.5543	0.4501	0.4216	0.4812
14	0.4662	0.3868	0.4177	0.3985
16	0.3103	0.3288	0.2288	0.1798
18	0.1514	0.1129	0.1218	0.1046
20	0.1654	0.1136	0.0622	0.0660
22	0.1040	0.0734	0.0527	0.0580

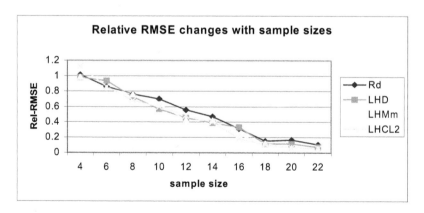

Table A.4. Two way ANOVA table

Source	SS	df	MS	F
Column	0.3062	3	0.1021	5.082
Rows	61.4	9	6.822	339.7
Interaction	0.7511	27	0.02782	1.385
Error	11.25	560	0.02008	
Total	73.7	599		

PV2 (p-values for the two-way ANOVA) follow: 0.0018, 0, 0.0950. The first number shows column (design) effect, the second row (sample size) effect, and the third interaction between the two. Since the first value is far below 0.01, the different design types do impact the accuracy. Even so, the sample size does have larger impact, not to mention the cases when there is no significant difference among the errors generated by different designs.

The following pie charts are generated from the ANOVA tables when using the first ten functions for comparing RandSamp, LHD, LHMm, and LHCL2, with n = 2, 5, and 10. The charts show clearly that the red color or run number effect occupies much larger portion of the total variation than the blue color or the design type effect. The yellow color shows the portion by interaction and within group (10 members/group) variation.

For most of the cases tested, it shows that enough samples are more important than "better designs" when Kriging models are used. Sometimes, adding only several runs might provide more accurate approximation than that by "better designs." Using enough samples might be more efficient than spending long time to find better designs if they exist and if the time for high fidelity model running does not take too long.

The sample size determination for metamodeling is likely to be related to the following factors as a starting point: the number of design variables; the characteristics of the approximation model; the number of parameters estimated for the model; the complexity of the underlying function (relationship); the radius of the hyper-sphere containing the design space; variable conversion considerations, normalization, the units used, related to the radius; desired prediction accuracy; and experimental design approach.

Many values in the third column of the p-values for two-way ANOVA are below 0.01, indicating there is significant interaction between sample sizes and design types in many cases.

Variation analysis based on two-way ANOVA tables
Comparing Rd LHD LHMm LHCL2

N=2, m=4, 6, ..., 22

AC

	SS	%
design	9.163	2.5
run no.	173.3	47.9
others	179.137	49.5
Total	361.6	100.0

MI

	SS	%
design	0.00509	0.5
run no.	0.3865	37.4
others	0.64241	62.1
Total	1.034	100.0

AX

	SS	%
design	1.03E+06	2.4
run no.	3.23E+07	75.1
others	9718000	22.6
Total	4.31E+07	100.0

BR

	SS	%
design	1257	0.4
run no.	2.51E+05	83.4
others	48443	16.1
Total	3.00E+05	100.0

DE

	SS	%
design	6.29E+05	3.3
run no.	1.47E+07	75.9
others	4010800	20.8
Total	1.93E+07	100.0

GD

	SS	%
design	1.80E+10	1.8
run no.	3.95E+11	40.4
others	5.65E+11	57.8
Total	9.78E+11	100.0

RB

	SS	%
design	7.03E+14	1.2
run no.	4.89E+16	80.8
others	1.09E+16	18.0
Total	6.05E+16	100.0

SX

	SS	%
design	1893	6.8
run no.	1.40E+04	50.1
others	12097	43.2
Total	2.80E+04	100.0

RY

	SS	%
design	1.48E+06	1.5
run no.	7.97E+07	78.7
others	20132000	19.9
Total	1.01E+08	100.0

PK

	SS	%
design	1.493	1.5
run no.	46.92	48.0
others	49.297	50.5
Total	97.71	100.0

N=5   m=5,10,…,50
AC

	SS	%
design	0.1268	1.0
run no.	3.804	30.9
others	8.3692	68.0
Total	12.3	100.0

AX

	SS	%
design	1.46E+06	2.1
run no.	3.33E+07	47.2
others	35859000	50.8
Total	7.06E+07	100.0

DE

	SS	%
design	9.09E+04	1.3
run no.	4.20E+06	61.6
others	2529090	37.1
Total	6.82E+06	100.0

RB

	SS	%
design	2.10E+14	0.8
run no.	1.34E+16	52.6
others	1.184E+16	46.6
Total	2.54E+16	100.0

RY

	SS	%
design	5.13E+06	0.6
run no.	6.09E+08	67.0
others	294975000	32.5
Total	9.09E+08	100.0

N=10   m=10, 20,30,…,100
AC

	SS	%
design	3.94E-03	2.6
run no.	2.87E-02	18.7
others	0.120665	78.7
Total	1.53E-01	100.0

AX

	SS	%
design	1.67E+06	5.9
run no.	3.50E+06	12.4
others	23032000	81.7
Total	2.82E+07	100.0

DE

	SS	%
design	1.36E+05	10.9
run no.	3.36E+05	27.1
others	768500	62.0
Total	1.24E+06	100.0

RB

	SS	%
design	5.11E+14	8.4
run no.	1.42E+15	23.3
others	4.143E+15	68.3
Total	6.07E+15	100.0

RY

	SS	%
design	5.33E+07	4.3
run no.	4.37E+08	35.0
others	757140000	60.7
Total	1.25E+09	100.0

PIE CHARTS (corresponding to the sheets above)

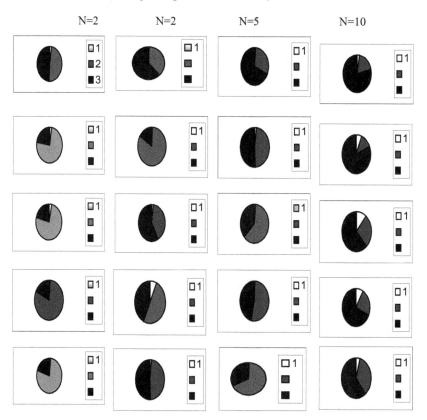

Figure A.6.2. The portions of the variation caused by designs, sample sizes, interaction plus within group variation

1:    The portion of the total variation caused by different design types
2:    The portion of the total variation caused by different run numbers,
      Increment: 2 (for n=2); 5 (for n=5, 10).
3:    The portion of the total variation caused by interaction between columns and rows, or designs and run numbers, plus within group variation.

A.7. Is more always better than less uniformly distributed sampling for computer experiments?

It has shown LHCL2 is often better than LHMm when there are significant differences between the means of the errors. The following matrix shows one example of the average RMSE and maximum error resulted from using LHMm and LHCL2. The test function is AX. The number of design variables is 2. The number of runs (m) is 4, 6, 8, 10, 12, 14, 16, 18, 20, 22.

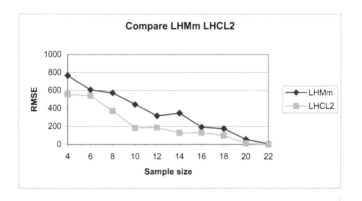

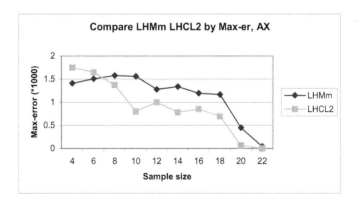

Figure A.7.1 Comparison between LHMm and LHCL2 based on RMSE and maximum error for function AX.

However, according to the author's simulation, LHMm almost always generates more or much more uniformly distributed samples in two dimension designs than those by LHCL2. It is generally agreed that the better designs should have more uniformly distributed sampling (see e.g. Santner, et al, 2003, p124). Why do the results here seem against the "general feeling?" For further investigation, many groups of twenty samplings from LHMm and LHCL2 each were tested. For each sampling, the twenty test functions were approximated with Kriging models. The error used for comparison is the relative root mean square error:

$$RMSE = \sqrt{\frac{1}{5000} \sum_{i=1}^{5000} \left(\frac{Ft(i) - Fa(i)}{|Ft(i)| + \varepsilon} \right)^2}$$

Ft: True response; Fa: approximation; ε: 10^{-6}: to guard against possible Ft = 0.

For verification, only one sampling group of 5000 points was generated which is used by all the tests. The one way ANOVA was applied to compare LHMm and LHCL2. This process was repeated for ten times (with the same n=2, m=2,4,...,22). The sampling plots and ANOVA plots together with the p-values are shown in Figures A.7.6 and A.7.7.

In most cases, there is no significant difference between the means of the errors generated by the two different designs. Although in many cases it is obvious that LHMm generates more or much more uniformly distributed sampling, the means of errors are not significantly different. In one case, LHMm is significantly worse than LHCL2 instead of being better as expected.

More cases were similarly tested for comparing LHMm with LHD, and LHMm with random sampling. Again, LHMm generates more or much more uniformly distributed sampling, yet no significant difference in the means of the errors is seen. In some cases, LHMm is worse than the other design despite its more uniformly distributed sampling. The averages of RMSE and the averages of maximum errors are shown below, each element being the average of the results of 20 functions. The sampling plots and ANOVA plots together with the p-values are shown in Figures A.7.8-11.

More similar tests were conducted. Occasionally, there is significant difference in the maximum error, with LHMm better or opposite, with more uniformly distributed LHMm sampling and other less uniformly distributed sampling by Rd, LHD, or LHCL2.

1) Comparing LHMm and LHCL2

Averages of RMSE=
```
0.2498   0.2139
0.1549   0.1917
0.1400   0.0840
0.1368   0.0535
0.0607   0.0386
0.0553   0.0418
0.0521   0.0401
0.0352   0.0311
0.0332   0.0250
0.0251   0.0230
```

Averages of Max-error=1.0e+003 *
```
2.3786   2.4459
0.6101   2.4450
0.1258   0.1077
0.1472   0.1126
0.1104   0.1113
0.1149   0.1354
0.1504   0.1040
0.1181   0.0974
0.1077   0.1210
0.1353   0.1087
```

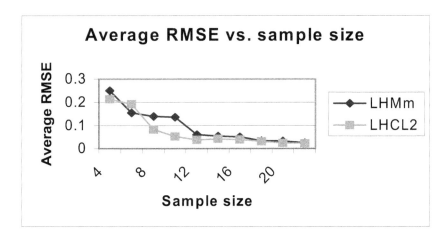

2) Comparing LHMm and LHD

Averages of RMSE=
 0.2498   0.2242
 0.1678   0.1882
 0.1204   0.0883
 0.1527   0.0831
 0.0632   0.0664
 0.1325   0.0347
 0.0397   0.0366
 0.0325   0.0347
 0.0290   0.0258
 0.0229   0.0317

Averages of Max-error=1.0e+003 *
 2.3786   2.4795
 2.3943   1.7958
 0.1191   0.1078
 0.1477   0.1276
 0.1128   0.1069
 0.1414   0.1152
 0.1289   0.1105
 0.1117   0.1205
 0.1279   0.1306
 0.1314   0.1600

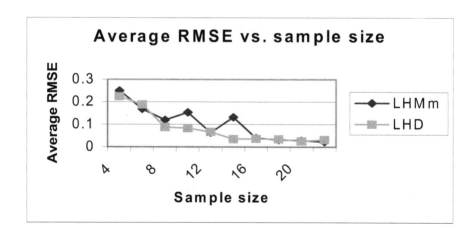

3) Comparing LHMm and Random sampling

Averages of RMSE=

0.2498	0.2559
0.1908	0.1822
0.1127	0.0919
0.1475	0.1158
0.0636	0.0716
0.0876	0.0544
0.0488	0.0445
0.0363	0.0340
0.0291	0.0352
0.0253	0.0283

Averages of Max-error

0.6014	0.7316
0.5653	0.6097
0.4471	0.4107
0.4950	0.4843
0.2907	0.3989
0.3692	0.3204
0.2856	0.2658
0.2365	0.2065
0.1909	0.1973
0.1734	0.1747

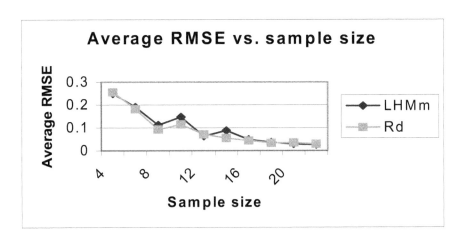

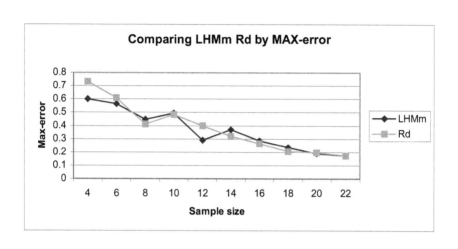

Figure A.7.6. Sampling plots and one-way ANOVA plots: Number 1-5, comparing RMSE means from LHMm and the means from LHCL2.

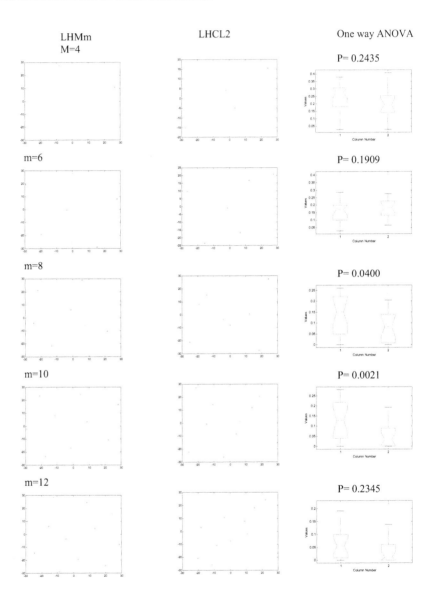

Figure A.7.7. Sampling plots and one-way ANOVA plots: Number 6-10, comparing RMSE by LHMm and LHCL2.

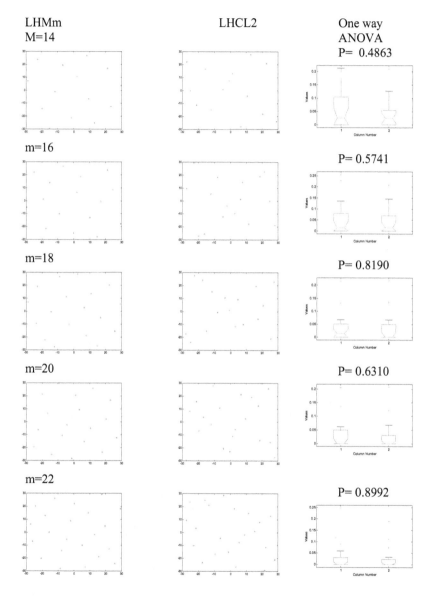

Figure A.7.8. Sampling plots and one-way ANOVA plots: Number 1-5, comparing RMSE by LHMm and LHD

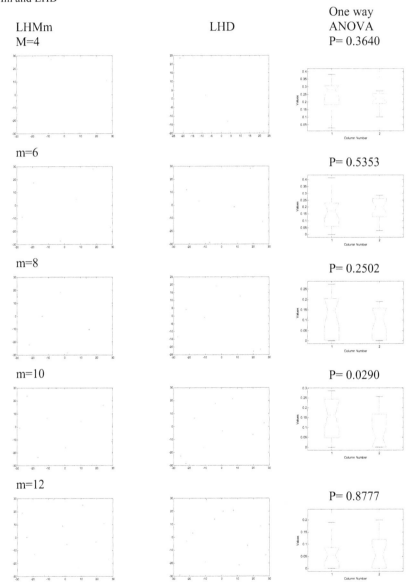

LHMm

LHD

One way
ANOVA

M=4

P= 0.3640

m=6

P= 0.5353

m=8

P= 0.2502

m=10

P= 0.0290

m=12

P= 0.8777

Figure A.7.9. Sampling plots and one-way ANOVA plots: Number 6-10, comparing RMSE by LHMm and LHD.

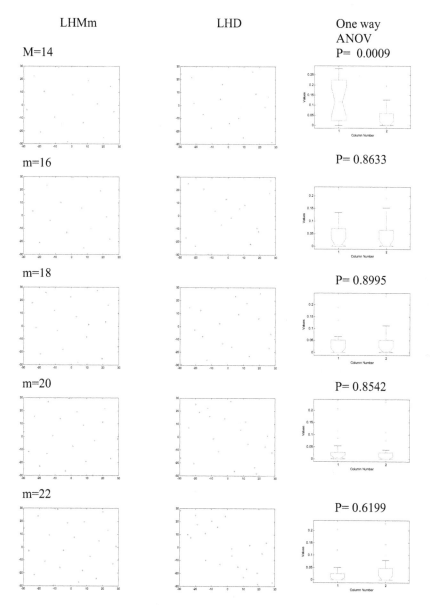

Figure A.7.10. Sampling plots and one-way ANOVA plots: Number 1-5, comparing RMSE by LHMm and Random sampling (Rd).

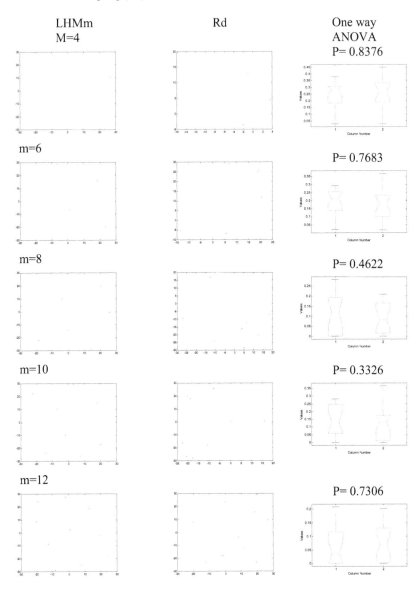

Figure A.7.11. Sampling plots and one-way ANOVA plots: Number 6-10, comparing RMSE by LHMm and Random sampling (Rd).

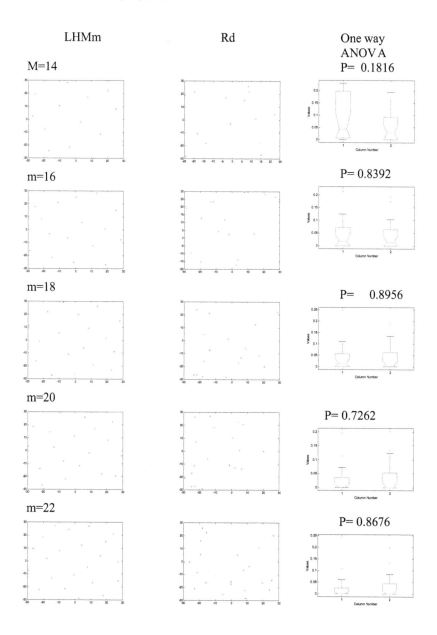

A.8.   Conclusions with limitations and discussion

Two approaches for global optimization are proposed for optimizing LHD. The stochastic search might be more effective and efficient when very long and complicated computation is needed by other approaches. "Standard" genetic algorithm approach proposed might be more effective than other approaches for the global optimization tasks, e.g. simulated annealing (however, the comparison study has not been done).

Due to the extremely limited resources for this study, limited simulation investigation on the effect of different design types based on different criteria or their combinations has been conducted. For the cases tested, the design types investigated often do not have large impact on the prediction accuracy. Among those cases that do see the significant differences, Latin hypercube sampling is mostly better than random sampling. Maximin LHD (LHMm) is often worse than LHD and LHCL2, but better sometimes in 2D cases when the sample size is close to 20 (the relative errors are often close to 0.01, 0.001 or lower). The Minimized CL2 Latin hypercube design (LHCL2) is often better than LHD and LHMm in 2D cases. In 5D and 10D cases, LHD appear better than the other designs; LHCL2 is mostly better than LHMm. The uniform designs based on the tables on the web can be better than any other designs for some cases, but much worse in some other cases, when the significant differences exist.

This study is a preliminary investigation for the topics studied. It needs comprehensive testing for making general conclusions through simulation. In this study, twenty test functions were tested. Many more are desired. The output from the computer programs might behave differently from the test functions, thus it may be needed to test those output. The numbers of design variables tested are 2, 5, and 10. Apparently, for large problems, many larger numbers need to be tested. The run number tested is up to 100. Higher numbers may be needed for large problems. And, only a portion of the design types in the literature was tested. The other types need be tested also. Only Kriging models were used; other models like support vector regression models and those generated by genetic programming can be tried beyond neural nets, radial basis functions, multivariate adaptive regression spline (MARS), polynomials, etc.

For those designs with combined criteria for optimizing LHD, the problems belong to multi-objective optimization (MOO) or MDO (multidisciplinary optimization) area. The methodologies in MOO or MDO might help to find lower combined criterion values. This work using MOO or MDO methodologies has not been tried. More study is needed to compare the designs with combined criteria and to compare them with the designs with single criterion. Pairwised comparison between any pair of the designs might be needed for refined comparison.

Although the testing range is limited, this study reveals that it may be more important to have enough samples than to have better designs in many cases. It is often necessary to use enough samples to reach desired error level. Using the same sample size, different designs may result in different accuracy. The detailed guidelines for sample size determination may be invaluable to the users of computer experiments. It is desirable that the users could be told about the tradeoffs between the effort (in sampling and running the program) and the result (accuracy level).

It is shown that LHD and LHCL2 is often better than LHMm for prediction accuracy in terms of root mean square errors (RMSE) when there are significant differences, including 2D cases. Yet, LHMm almost always generates more uniformly distributed sampling in 2D. It is clear that more is not necessarily better than less uniformly distributed sampling for reducing RMSE. Much more study on uniformity effect is needed.

Among the designs tested, LHCL2 or LHD is recommended for the 2D cases similar to those tested; LHD is recommended for the 5D and 10D cases similar to those tested, with less sampling effort but better prediction results.

REFERENCES FOR APPENDIX

Barton, R. R., 1994, December 11-14, "Metamodeling: A State of the Art Review," *Proceedings of the 1994 Winter Simulation Conference (Tew, J. D., Manivannan, S., et al., eds.),* Lake Beuna Vista, FL, IEEE, pp. 237-244.

Barton, R. R., 1998, December 13-16, "Simulation Metamodels," *Proceedings of the 1998 Winter Simulation Conference (WSC'98) (Medeiros, D. J., Watson, E. F., et al., eds.),* Washington, DC, IEEE, pp. 167-174.

Barton, R. R. 1992. Metamodels for simulation input- output relations, Proceedings of the 1992 Winter Simulation Conference, ed. D. Goldsman, J. Swain, J. Wilson, 41 1-41 8. IEEE, Piscataway, New Jersey.

Chipperfield, A., Fleming, P., Pohlheim, H., and Fonseca, C. *Genetic Algorithm Toolbox: User's guide,* http://www.shef.ac.uk/~gaipp/ga-toolbox/manual.pdf

Fang, K. T., 1980, "Experimental Design By Uniform Distribution," *Acta Mathematice Applicatae Sinica,* Vol. 3, No. 363-372, pp.

Fang, K.-T. and Wang, Y., 1994, *Number-theoretic Methods in Statistics,* Chapman & Hall, New York.

Fang, K.T., Lin, D.K.J., Winker, P. and Zhang, Y. (2000), Uniform design: Theory and applications, *Technometrics,* v42, n3, pp237-248

Fang, K.T., Lin, D., and Liu, M., 2000, Optimal mixed-level supersaturated design and computer experiment, http://citeseer.ist.psu.edu/cache/papers/cs/17273/ftp

Fang, K.T. , Ma, C.X. and Winker, P. (2002): Centered L2-discrepancy of random sampling and Latin hypercube design, and construction of uniform designs, *Mathematics of Computation,* 71, 275-296.

Goldberg D.E. (1989), *Genetic algorithms in search, optimization and machine learning,* Addison-Wesley Publishing Co., Inc., Reading, Massachusetts.

Hickernell, F.J. (1998): A generalized discrepancy and quadrature error bound, *Mathematics of Computation,* 67, 299-322.

Holland J. (1975), *Adaptation in natural and artificial systems,* University of Michigan Press, Ann Arbor.

Iman, R.L. and Conover, W.J. (1982): A distribution-free approach to inducing rank correlation among input variables, *Communications in Statistics ,* Series B 11,311-334.

Jin, R., Chen W., and Sudjianto, A. (2003): An efficient algorithm for constructing optimal design of computer experiments, ASME 2003 Design engineering technical conferences, Chicago, Illinois, USA, Sept. 2-6.

Johnson, M. E., Moore, L. M. and Ylvisaker, D., "Minimax and Maximin Distance Designs," *Journal of Statistical Planning and Inference,* Vol. 26, No. 2, 1990, pp. 131-148.

Koehler, J. R. , Owen, A. B. 1996: *Computer experiments.* In Ghosh, S. and Rao, C. R., editors, *Handbook of Statistics*, **13**, 261--308. Elsevier Science, New York.

Liefvendahl, M. and Stocki, R: A study on algorithms for optimization of Latin hypercubes, http://www.ippt.gov.pl/~rstocki/JSPI-olh.pdf

Li, W., and Wu,C.F.J. (1997): Columnwise-Pairwise algorithms with applications to the constructionof supersaturated designs, *Technometrics*, 39, 171-179.

McKay, M.D., Beckman R.J. and Conover W.J. 1979: A comparison of three methods for selecting values of input variables in the analysis of output from a computer code. *Technometrics*. **21**, 239-245 (reprinted in 2000: *Technometrics*, **42**, 55-61.

Morris, M. D. and Mitchell, T. J., 1995, "Exploratory Designs for Computer Experiments," *Journal of Statistical Planning and Inference*, 43, 381-402.

Nielsen, H.B., Lophaven, S.N., and Søndergaard, J.:   DACE: A   MATLAB   Kriging Toolbox, http://www.imm.dtu.dk/~hbn/dace/

Osio, I. G. and Amon, C. H., "An Engineering Design Methodology with Multistage Bayesian Surrogates and Optimal Sampling," *Research in Engineering Design,* Vol. 8, No. 4, 1996, pp. 189-206.

Owen, A. B., 1992, "Orthogonal Arrays for Computer Experiments, Integration and Visualization," *Statistica Sinica,* Vol. 2, pp. 439-452.

Owen, A.B. (1994): Controlling correlations in latin hypercube samples, *Journal of the American Statistical Association*, 89, 1517-1522.

Palmer, K. et al., "Minimum bias Latin hypercube design, " IIE Transactions, 2001, 33, 793-808.

Park, J.-S., 1994, "Optimal Latin-Hypercube Designs for Computer Experiments," *Journal of Statistical Planning and Inference,* Vol. 39, No. 1, pp. 95-111.

Sacks, J., Welch, W.J., Mitchell, T.J. and Wynn, H.P. (1989). "Design and Analysis of Computer Experiments." *Statistical Science* **4**(4): 409-435.

Sacks, J., Schiller, S.B. and Welch, W.J., 1989, "Designs for Computer Experiments," *Technometrics,* Vol. 31(1), pp. 41-47.

Sahama, T. and Diamond, N.T. (2001): Sample size considerations and augmentation of computer experiments, *Journal of Statistical computation and simulation*, 68, 307-319.

Santner, T., Williams, B, Notz, W. 2003: *Design and Analysis of Computer Experiments,* 150, Springer-Verlag, New York.

Shewry, M.C. and Wynn, H.P. (1987): Maximum entropy sampling, *Journal of Applied Statistics*, 14, 165-170.

Simpson, T. W., Mauery, T. M., Korte, J. J. and Mistree, F., "Kriging Metamodels for Global Approximation in Simulation-Based Multidisciplinary Design Optimization," *AIAA Journal,* Vol. 39, No. 12, 2001, pp. 2233-2241.

Simpson, T. W., Booker, A. J., Ghosh, D., Giunta, A. A., Koch, P. N., and Yang, R.-J. "Approximation Methods in Multidisciplinary Analysis and Optimization: A Panel Discussion," *Structural and Multidisciplinary Optimization,*

Simpson, T. W., Peplinski, J., Koch, P. N. and Allen, J. K. (2001) "Metamodels for Computer-Based Engineering Design: Survey and Recommendations," *Engineering with Computers*, **17**:2 (129-150)

Tang, B., 1993, "Orthogonal Array-Based Latin Hypercubes," *Journal of the American Statistical Association,* Vol. 88, No. 424, pp. 1392-1397.

Tang, B., 1994, "A Theorem for Selecting OA-Based Latin Hypercubes Using a Distance Criterion," *Communications in Statistics, Theory and Methods,* Vol. 23, No. 7, pp. 2047-2058.

Welch, W. J., Buck, R. J., Sacks, J., Wynn, H. P., Mitchell, T. J. and Morris, M. D., 1992, "Screening, Predicting, and Computer Experiments," *Technometrics,* Vol. 34, No. 1, pp. 15-25.

Welch, W. J., Yu, T.-K., Kang, S. M. and Sacks, J., "Computer Experiments for Quality Control by Parameter Design," *Journal of Quality Technology,* Vol. 22, No. 1, 1990, pp. 15-22.

Appendix II

Sample size, uniformity, and design method comparison for computer experiments

(2005, newly added for this publication)

Sample Size, Uniformity, and Design Method Comparison for Computer Experiments

ABSTRACT

Is having enough samples more important than using better designs for computer experiments? Does more uniform sampling generally lead to more accurate approximation? Which of the different criteria used to optimize Latin hypercube designs (LHDs) yields the best design for prediction purpose? This study compared four designs extensively and fourteen more designs briefly by evaluating approximation errors. The differences in error means were often not statistically significant. When they were significant, LHD was mostly better than random sampling (Rd); Maximin LHD (LHMm) was sometimes better than other designs, but often worse especially in 5D and 10D cases; Minimized-CL2 LHD (LHCL2) was often better than LHD and LHMm in 2D cases and was similar to LHD in 5D and 10D cases. Sample sizes often had stronger impact on the accuracy than design types, and enough samples were needed to achieve low error levels when using one-stage sampling. More uniform sampling did not generally lead to more accurate prediction unless sampling included extremely nonuniform cases.

Key words: sample size, uniformity, experimental design, computer experiments, metamodeling

d_{ij} the Euclidean distance between points i and j

n (N)   the number of design variables

m (M)   the number of runs or sample size

q   a parameter to be determined

CL_2 Centered L_2 discrepancy

LHCL2   Minimized-CL2 Latin hypercube design

LHD   Latin hypercube design

LHMm   Maximin Latin hypercube design

ML_2 Modified L_2 discrepancy

PS   Proportional sampling

R, R1   weighting parameters

Rd   Random design

SL_2 Symmetric L_2 discrepancy

X_{ij} the jth component of the ith sampled point

U_{ij} the jth element of ith independent random variable U [0, 1] (uniform distribution in [0,1]), independent of the π_{ij}

Φ_q a design criterion

π_{ij} the jth element of ith independent uniform random permutations of the integers 1 through n

1. INTRODUCTION

Latin hypercube designs (LHDs) have been widely used for compuier experiments in many fields. Their projections onto one-dimensional subspaces are evenly distributed and are relatively easy to generate (Santner, et al, 2003). They do not, however, always produce space-filling designs.

Many criteria have been proposed and employed for generating space-filling designs. To name a few, Shewry and Wynn (1987) introduced maximum entropy sampling. Johnson et al (1990) developed minimax and maximin designs. Owen (1992,1994) and Tang (1992, 1993) considered using orthogonal arrays. Fang and Wang (1994) introduced uniform designs.

Many criteria produce experimental designs with attractive properties, but none of them alone can be completely satisfactory. It would be ideal if one approach could offer all of the desired features of these criteria. One way to find such an approach would be to generate many LHDs, and then to select the design with the best criterion or combined criterion value. The single or combined criterion that consistently results in high accuracy would then be considered the ideal criterion.

Several approaches have been proposed to optimize LHD. Morris and Mitchell (1995) extended maximin distance criterion and introduced a Φ_q criterion. Fang, et al, (2002) optimized LHD by minimizing centered discrepancy CL2. Iman and Conover (1982) and Owen (1994) minimized a linear correlation criterion for pairwise factors. Tang (1993) selected LHD that was orthogonal after some transformation. For other approaches, see for example, Santner, et al (2003).

While comparing different approaches for experimental designs or optimizing LHDs, many researchers compare the designs based on the criterion values rather than how the designs serve the goals for computer experiments such as prediction. That may be problematic. For instance, some people use maximin criterion for evaluating designs. It is well known, and confirmed by our simulation tests, that maximin designs tend to place points near the boundary of the sampling region (Santner, et al, 2003). Thus the designs are not necessarily space-filling designs, nor do they necessarily result in more accurate approximation.

As one of the major goals for computer experiments is prediction, the main research questions addressed in this investigation included: 1) How do different designs perform in this regard? 2) Would it be better to use other criteria than Φ_q criterion and CL2 to optimize LHD (e.g. modified L_2 discrepancy ML2, symmetric L_2 discrepancy SL2, column-wise correlation for column orthogonal design, etc., or combinations of these criteria)? 3) How do sampling size and uniformity affect prediction accuracy?

To optimize LHD, several optimization approaches had been suggested. Many researchers used simulated annealing (SA) or its variations, e.g. Morris and Mitchell (1995), Fang, et al (2002), Park (1994), Li and Wu (1997), Jin, et al (2003), etc. For global optimization, genetic algorithms (GA) had been successfully applied in many fields. We used GA to optimize the criteria, and found it to be efficient. The details (different from those of

Liefvendahl and Stocki) are available upon request. Since GA and other methodologies for optimizing LHD may involve many calculations and significant memory space, we also tested an extremely simple approach— random search. It requires minimal calculation, needs much less memory, and outperformed GA in some cases with less computation and better results.

The layout of this paper is as follows. The different design types under investigation are described. The comparison begins with the four primary types: Rd, LHD, LHMm, and LHCL2, followed by all eighteen types. The impact of sample sizes and sampling uniformity on prediction accuracy is then discussed. Test results with some very nonuniform samplings are presented to stress the need to check extremely nonuniform samplings. A discussion section providing conclusions and limitations completes the paper.

2. THE EIGHTEEN DESIGN TYPES AND TWENTY TEST FUNCTIONS

Eighteen design types

1) Random design (Rd)

2) Latin hypercube design (LHD) (McKay, et al, 1979): The jth component of the ith sampled point is

$$X_{ij} = \frac{\pi_{ij} - U_{ij}}{m} \tag{1}$$

3) Maximin Latin hypercube design (LHMm)

Criterion: Φ_q criterion (slightly modified in the form from that of Morris and Mitchell, 1995)

$$\Phi_q = \left[\sum_{i=1}^{m} \sum_{j=i+1}^{n} d_{ij}^{-q} \right]^{1/q} \tag{2}$$

q: Pilot tests show better values: 1 for most 2D cases, 75 or 45 for 5D cases, 75 for most 10D cases; testing with q = 2 or other values were also conducted with similar or a little worse performance.

4) Minimized CL2 Latin hypercube design (LHCL2)

Criterion: Centered L_2 discrepancy CL_2 (Hickernell, 1998)

$$[CL_2(P_m)]^2 = (\tfrac{13}{12})^2 - \tfrac{2}{n} \sum_{k=1}^{m} \prod_{j=1}^{n} [1 + \tfrac{1}{2}|x_{kj} - 0.5| - \tfrac{1}{2}|x_{kj} - 0.5|^2]$$
$$+ \tfrac{1}{n^2} \sum_{k=1}^{m} \sum_{j=1}^{m} \prod_{i=1}^{n} [1 + \tfrac{1}{2}|x_{ki} - 0.5| + \tfrac{1}{2}|x_{ji} - 0.5| - \tfrac{1}{2}|x_{ki} - x_{ji}|] \tag{3}$$

5) Minimized ML2 Latin hypercube design (LHML2)

Criterion: Modified L$_2$ discrepancy ML$_2$ (Hickernell, 1998)

$$[ML_2(P_m)]^2 = (\tfrac{4}{3})^n - \frac{2^{1-n}}{m} \sum_{k=1}^{m} \prod_{l=1}^{n} (3 - x_{kl}^2) + \frac{1}{n^2} \sum_{k=1}^{m} \sum_{j=1}^{m} \prod_{i=1}^{n} [2 - \max(x_{ki}, x_{ji})] \qquad (4)$$

6) Minimized SL2 Latin hypercube design (LHSL2)

Criterion: Symmetric L$_2$ discrepancy SL$_2$ (Hickernell, 1998)

$$[SL_2(P_m)]^2 = \left(\frac{4}{3}\right)^n - \frac{2}{m} \sum_{k=1}^{m} \prod_{j=1}^{n} (1 + 2x_{kj} - 2x_{kj}^2) + \frac{2^n}{m^2} \sum_{k=1}^{m} \sum_{j=1}^{m} \prod_{i=1}^{n} [1 - |x_{kj} - x_{ji}|] \qquad (5)$$

7) Column orthogonal Latin hypercube design (LHOrth)

Criterion: Orth − maximum value of the correlation coefficients between two different

columns in the design matrix

8) Maximin CL2 LHD (LHMmCL2), Criterion: Φ_q + R * CL2

9) Maximin ML2 LHD (LHMmML2), Criterion: Φ_q + R * ML2

10) Maximin SL2 LHD (LHMmSL2), Criterion: Φ_q + R * SL2

11) Maximin column orthogonal LHD (LHMmOrth), Criterion: Φ_q + R * Orth

12) CL2 column orthogonal LHD (LHCL2Orth), Criterion: CL2 + R * Orth

13) ML2 column orthogonal LHD (LHML2Orth), Criterion: ML2 + R * Orth

14) SL2 column orthogonal LHD (LHSL2Orth), Criterion: SL2 + R * Orth

15) Maximin CL2 column orthogonal LHD (LHMmCL2Orth)

Criterion: Φ_q + R * CL2 + R1 * Orth

16) Maximin ML2 column orthogonal LHD (LHMmML2Orth)

Criterion: Φ_q + R * ML2 + R1 * Orth

17) Maximin SL2 column orthogonal LHD (LHMmSL2Orth)

Criterion: $\Phi_q + R * SL2 + R1 * Orth$

18) Uniform designs (UD) (Centered $L_2 U_n(n^s)$ Tables located at

http://www.math.hkbu.edu.hk/UniformDesign)

Twenty test functions

Most of the functions tested are popular functions for testing global optimization methodologies. Many of them have high nonlinearity or multi-modes, whereas others are quite simple or smooth. The details of the functions are presented in the Appendix.

3. COMPARISON OF 18 DESIGN TYPES FOR PREDICTION ACCURACY

3.1 Test scheme

The main study tools employed were simulation tests and statistical analyses based on one-way and two-way ANOVA (analysis of variance). After a test function and a sample size were selected, an experimental design method was employed for sampling. Ten (in 5D and 10D cases) or 20 (in 2D cases) sampling groups were generated as 10 or 20 "observations" since they were not deterministic samplings. For each group, a kriging model was used to approximate the test function. The validation sites were generated by Latin hypercube design to provide 5000 (in 5D or 10D cases) or 10,000 (in 2D cases) points within the design space. The sole validation set was used by all the samplings for the same function and the same dimension. The true values and those evaluated by kriging models were compared to find root mean square errors (RMSE), maximum errors, or relative RMSE (RMSE divided by the mean of all the responses for validation). The above process was then repeated for each design type. Next, the sample size was increased and the process was repeated. Finally, the entire process was repeated for another function. The numbers of dimensions tested were 2, 5, and 10. The sample sizes varied from 4 to 300.

One way and two-way ANOVA were employed to compare different designs by reviewing the means of RMSE to see any statistically significant differences as well as evaluating the range of error distribution. The only factor for one-way ANOVA was design type. The two factors for two-way ANOVA included a column factor corresponding to design type and a row factor according to sample size. In some cases, the means of the maximum errors were compared as well. The critical p-value was selected as 0.01 after comparing several p-values, F values and ANOVA plots. Two of the one-way ANOVA plots with p-values close to 0.01 and 0.05 are shown in Fig. 1. The test results with 5% were also shown.

The above testing and ANOVA comparisons were repeated for each test function separately. There was no mixed comparison between different test functions in this section. For one-way ANOVA, the explanation of the box plots follows. The centerline is the median; the bottom and top lines of the box are the quartiles; the end lines of the whiskers are the ends of the data range. The outliners are those marked as "+". The dots on the end lines show that there is no outliner.

3.2. Comparison among Rd, LHD, LHMm, LHCL2 by GA

It was desirable though not practical to comprehensively compare all of the design types under investigation. Accordingly, only four of the design types, Rd, LHD, LHMm, LHCL2, underwent extensive testing as is shown below. All twenty test functions were used for 2D cases, and the first five functions (AC, AX, DE, RB, RY) were used for 5D and 10D cases.

1) Two design variable cases (n = 2).
The sample sizes ranged from 4 to 22. Among all 200 cases (there were 20 sampling groups for each sample size), 86 cases or 43% of the total cases had one-way ANOVA p-values below 0.01, and 109 cases (54.5%) below 0.05. In most of these cases, LHD was

better than Rd; LHMm was sometimes better than Rd and LHD. LHCL2 was mostly better than Rd and LHD, often better than LHMm. Fig. 2 shows the average RMSE vs. sample sizes when comparing the four designs simultaneously.

Since Rd and LHD behaved so similarly, pairwise tests were conducted to get the exact number of the cases that LHD was significantly different from Rd. Similarly, pairwise tests LHD vs. LHMm, LHD vs. LHCL2, and LHMm vs. LHCL2 were conducted.

Among 200 cases when Rd and LHD were compared, the p-values for one-way ANOVA in 17 cases (8.5%) were below 0.01. In all the cases, LHD was significantly better than Rd. When 0.05 was used as the critical value, 48 cases (24%) had p-values below 0.05. LHD was better than Rd in 46 cases. Among 200 cases when LHD and LHMm were compared, 54 cases (27%) had significant differences (p-v < 0.01). In 27 cases, LHD was better; in 27 cases, LHMm was better. Among 200 cases when LHD and LHCL2 were compared, 17 cases (8.5%) had significant differences. In only 2 cases, was LHD better; in 15 cases, LHCL2 was better. Among 200 cases when LHMm and LHCL2 were compared, 78 cases (39%) had significant differences. In 30 cases, LHMm was better (mostly happened when the sample size was between 18 and 22); in 48 cases, LHCL2 was better (including some cases when the sample size was between 18 and 22).

2) Five design variable cases (n = 5).

The sample sizes ranged from 5 to 50 (extended to 100 in some comparisons). Among all 50 cases, 7 or 14% of the total cases had p-values below 0.01; 12 cases (24%) had p-values below 0.05. The results for pairwise tests follow. Among 50 cases when Rd and LHD were compared, 1 case had significant difference. LHD was better. Among 50 cases when LHD and LHMm were compared, 5 cases had significant differences. In all these cases, LHD was better. Among 50 cases when LHD and LHCL2 were compared, 4 cases had significant differences. LHD was better than LHCL2 in 3 cases and worse in the other case. Among 50 cases when LHMm and LHCL2 were compared, 6 cases had significant differences. In one case, LHMm was better; but in 5 cases, LHCL2 was much better. Fig. 5 shows the average RMSE vs. sample sizes when comparing the four designs simultaneously.

3) Ten design variable cases (n = 10).

The sample sizes ranged from 10 to 100 (extended to 300 in some comparisons). There were 13 cases (26%) showing p-values below 0.01; 21 cases (42%) below 0.05. The details of the comparison are shown in Table 1. Fig. 3 shows the average RMSE vs. sample sizes (10-100) when comparing the four designs simultaneously. Fig. 4 shows the error when comparing Rd and LHD for m = 10-300. Table 1 summarizes all the results including those for n = 2, 5, and 10.

3.3. Comparison among 18 design types

To evaluate 15 approaches for optimizing Latin hypercube designs together with RD, LHD, and uniform design, 18 design types were compared with one-way and two-way

ANOVA. Only the first ten functions were tested. Only 2D cases are reported here, with the sample sizes m = 4, 6, ..., 22. The number of points for validation was 5000, generated by LHD. Out of the total 100 cases, 37 cases (37%) showed significant differences (p-value < 0.01) in the means of the approximation errors when employing different designs; 60 cases (60%) showed p-values below 0.05. The means of RMSE (means of the 10 groups with the same dimension and sample size) vs. sample sizes are shown in Fig. 6. More study is needed.

4. WOULD ENOUGH SAMPLES BE MORE IMPORTANT THAN BETTER DESIGNS?

In many cases tested, there was no statistically significant difference (p-value > 0.01) between the approximation error means when using different design types. Even in those cases where significant differences existed, adding more runs led to higher accuracy than that by "better designs." The RMSE results indicated that there were larger differences among the results from different sample sizes than those from different designs. The two-way ANOVA tables revealed that the portion of the variation in terms of the sum of the squares due to different sample sizes was often much larger than that due to different designs. Table 2, Fig. 7, and ANOVA table (Table 3) show the cases for Branins' rcos function, while comparing Rd, LHD, LHMm, and LHCL2, with 2 design variables. Each element is the average of 20 output values (the relative RMSE: RMSE divided by the mean of the response values within the same validation group). The numbers are similar across the columns but not across the rows. It shows that the sample size has larger impact on RMSE means. Similar figures were shown previously for 2D, 5D, and 10D cases.

The p-values for the two-way ANOVA were 0.0018, 0, 0.0950. The first number refers to the design effect, the second to the sample size effect, and the third to interaction between the two. Both effects were significant. The first column, however, showed that the design type contributed 0.4% of the total variation while the sample size contributed 83.4%.

Table 4 was generated from the two-way ANOVA tables when using the first ten functions for comparing Rd, LHD, LHMm, and LHCL2, with n = 2, 5, and 10. The table showed clearly that the sample size effect explained a much larger portion of the total variation (in terms of the sum of squares) than did the design type effect in most cases. The third portion was caused by interaction and within group variation (10 or 20 members per group).

The test results showed that enough samples were likely to be more important than "better designs" when kriging models were used. In many cases, adding only two more runs reduced more error than using the "best" design for a given number of runs. Using enough samples might be more efficient than spending time trying to find better designs, assuming they exist and that the time to run a higher fidelity model (in metamodeling cases) is not excessive. In most cases tested, enough samples were necessary regardless of what design was used among those designs compared.

The sample size determination for computer experiments is likely to be related to the following factors as a starting point: the number of design variables; the characteristics of the approximation model; the number of parameters estimated for the model; the complexity of the underlying function (relationship); the radius of the hyper-sphere containing the design space; variable conversion considerations, normalization, and the units used; desired prediction accuracy; and experimental design approaches. Many values in the third column of the p-values for two-way ANOVA were below 0.01, indicating that in many cases there was significant interaction between sample size and design type.

206

Although the range of this study was limited, many scholars (see Acknowledgement for the names) provided similar observations in their comments on earlier manuscripts, regarding sample size effect.

5. DOES MORE UNIFORM SAMPLING GENERALLY LEAD TO MORE ACCURATE MODELING?

It has been shown that LHCL2 was often better than LHMm when there were significant differences between the means of the errors. Fig. 8 shows one example of the average RMSE and maximum error that resulted from using LHMm and LHCL2 for sampling. The test function was AX. The number of design variables was 2. The sample sizes were 4, 6, 8, 10, 12, 14, 16, 18, 20, and 22. This result and many other results seem surprising. According to our simulation, LHMm almost always generates more uniformly distributed samples in 2D and 3D designs than those by LHCL2. It is generally agreed that, by intuition, sampling for computer experiments should be uniformly distributed in the design region (e.g. Koehler and Owen, 1996; Santner, et al, 2003). The more uniformly distributed the sampling, the better the experimental design. Since uniformity has been taken as one of the fundamental issues for experimental design in computer experiments, and thus has significant impact on research and application, more tests were conducted for further investigation and are described below.

5.1. Test scheme

There were two test groups. In the first group, the four designs (LHMm, Rd, LHD, and LHCL2) were employed and compared simultaneously. Twenty test functions were approximated using each specific sampling, and there were five levels of the sample size.

The second group focused on one function, one level of sample size, and pairwise comparison each time.

In the first group, for the same sample size, the four designs were employed for sampling as four different "treatments." For each "treatment" that is a specific sampling, there were 20 "observations" that corresponded to 20 functions. For each function, a kriging model was built as the approximation model to fit these sampled points and the responses. To test the prediction accuracy, 10,000 validation points were generated by LHD within the domain [-30,30; -30,30]. The same domain was also used for sampling to build the models. The domain is specified in the literature for some of the test functions and was applied to all the functions in this section. At each point, the values of the test function and of the approximation model were compared to find the difference. From all the validation points, a relative error called relative root mean square error (RRMSE) was found by the following formula:

$$RRMSE = \sqrt{\frac{1}{10000} \sum_{i=1}^{10000} \left(\frac{Ft(i) - Fa(i)}{|Ft(i)| + \varepsilon} \right)^2} \qquad (6)$$

Ft: function response; Fa: approximation response; $\varepsilon = 10^{-4}$: to guard against possible Ft = 0.

And, the maximum relative error was found by the following formula:

$$Max - rel - error = Max(\frac{|Ft(i) - Fa(i)|}{|Ft(i)| + \varepsilon}), \qquad i = 1,2,...,10000 \qquad (7)$$

With exactly the same sampling, nineteen more RRMSE observations and nineteen more Max-rel-error observations were generated by the same procedure for approximating nineteen more test functions. This process was repeated for the other three design methods or "treatments." In all, there were twenty observations per each specific sampling to compare RRMSE and Max-rel-error respectively. One-way ANOVA was used to compare the designs by comparing the means of RRMSE and Max-rel-error respectively. The whole process outlined above was repeated for another sample size.

The sample sizes tested were 8, 10, 12, 14, and 16 to allow for visual comparison of the distribution uniformity. With too few or too many points, it was difficult to tell if the distributions were uniform by visual inspection. The relative error was employed because it was likely to be more meaningful for prediction. Further, since twenty test functions were used, using RMSE might tend to inflate within group variance, making it difficult to detect small differences in the error means resulted from different "treatments."

In the second group, many pairwise comparison tests between LHMm and one of the other three designs were conducted for one test function each time. After sampling, a kriging model was built and RMSE was calculated based on 10,000 validation points. The same process was repeated four more times for the same design type and the same sample size. Then, the whole process was repeated for the other design type. Next, ANOVA was conducted with each treatment having five observations of RMSE. Finally, the whole process was repeated on another pair of designs and another function.

5.2. Measure for uniformity

There seems to be no generally agreed upon definition for sampling uniformity. Many criteria have been proposed for reaching space filling or uniformly distributed designs, e.g. maximin, Fi (Φ_q) criterion, several discrepancy criteria, etc., as cited before. We conducted many 2D and 3D tests with LHMm. Those tests showed that LHMm resulted in more uniformly distributed samplings than other Latin hypercube designs. Thus, for the purpose of comparing the sampling uniformity of different designs, the Fi criterion was used as an indirect measure for uniformity, as follows. The measure is called relative uniformity (UNIF) shown by Eq. 8. The larger the UNIF, the larger the sum of the distances between the points, which results in a more uniform distribution within a stratified sampling domain such as LHD. It also seemed to be a good measure for Rd as well. More study is needed for the definition of (relative) uniformity.

$$UNIF(i) = FIinv(i) / MaxFIinv \tag{8}$$

UNIF: relative uniformity; FIinv (i): i[th] inversed Fi value corresponding to i[th] design;
MaxFIinv: the maximum inversed Fi value among those of the designs under comparison.

5.3. Results of the simulation tests and ANOVA

1) Two-dimension cases

a) Comparing four designs simultaneously

The sample sizes tested were: 8, 10, 12, 14, and 16. For validation, only one sampling group of 10,000 points was generated which was used by all the tests. The sampling plots and the ANOVA plots together with the p-values are shown in Figures 9-13.

b) Pairwise comparison between LHMm and Rd

The pairwise comparison between LHMm and one of the other three designs showed that again, in most cases, there was no significant difference in the RMSE means resulting from using LHMm versus another design. To show less uniformly distributed sampling can sometimes result in higher prediction accuracy, one pairwise comparison between LHMm and Rd is provided in Fig. 14. The test function was AC. The sample size was 10. In Rd sampling, the points cluster around one corner one time, around another corner at another time. In spite of this, "bad" Rd sampling still outperformed "better" LHMm sampling.

2) Three-dimension cases

The sample sizes tested were 12, 15, 18, 21, and 24. For validation, only one sampling group of 10,000 points was generated which was used by all the tests. The sampling plots, shown as the projections onto XY, XZ, ZY planes, and the ANOVA plots together with the p-values are shown in Figures 15-19.

It has been shown that LHMm generated more uniformly distributed sampling than did RD, LHD, or LHCL2. In most cases, however, there was no statistically significant difference in the means of RRMSE, Max-rel-error, or RMSE, regardless of the critical p-value selected as 0.01 or 0.05. In some cases, most of the points were along one straight

212

line or cluster in a corner, but still there was no significant difference in the RRMSE means. Occasionally, LHMm was significantly worse than Rd despite its being more uniformly distributed. Note that in the first group, twenty different functions, instead of only one function, were approximated using exactly the same specific sampling.

In terms of the variance or the range of the data (either RRMSE or Max-rel-error), it seemed that uniformity did not have consistent impact either, but sample size did matter. As the sample size increased, the range usually decreased, as shown in Figures 9-13. Note that the figures have different axis limits for the errors.

6.   SOME EXTREME CASES WITH A VERY NONUNIFORM
SAMPLING—PROPORTIONAL SAMPLING (PS)

Santner, Williams, and Notz (2003) described a sampling along a straight line, resulting in higher accuracy at the line but poor prediction elsewhere. It is possible for Rd or LHD to end up with some very nonuniform designs like this one. We refer to such a sampling as "proportional sampling" (PS) — one in which all sample points uniformly distribute along the diagonal line of the space. Some comparison results among Rd, LHD, LHMm, LHCL2, and PS are shown in Fig. 20. The test scheme was the same as that mentioned in Section 3 for comparing the first four design types. The results for uniformity testing, using the test scheme described in Section 5, but with PS replacing Rd, are presented in Figures 21-25.

When the sample size was relatively small, increasing the size resulted in a lower error level even with PS. When the size increased, other designs produced more widely spread samplings and decreasing levels of errors, but PS stayed along the diagonal line and kept the error level almost unchanged or even higher in many cases. In the latter cases when the error level increased as the sample size increased, the kriging models were misled by "over-fitting." It is shown in ANOVA plots that, when the size was relatively large, the error range by PS was much larger and the error level was higher than those by other design types. The results for even larger sizes (not shown) presented larger differences.

It is clear that very nonuniform designs should be avoided. Since it is possible for Rd and LHD to create such extreme samples, it is necessary to check the resulting design to guard against such extreme cases. The authors evaluated the Fi criterion as a possible measure for performing this check. When its inverse FIINV was very low, the sampling was quite nonuniform. The FIINV values for sampling are shown in Table 5.

In most cases, FIINV values by PS were lower than those by other design types. Unfortunately, FIINV became small when sample size was larger. After some experimentation, it was discovered that FIINV$*$m$^{2.5}$ resulted in relatively stable values vs. sample sizes, as is shown in Table 6. It appears that SFIINV = FIINV$*$m$^{2.5}$ might be a better tentative measure for "absolute" uniformity (vs. UNIF as described in Section 5).

In Table 6, the SFIINV values by PS were much lower than most other values by other design types. When SFIINV is quite low, another LHD can be generated and checked, until a higher SFIINV is obtained. Based on our simulation, the likelihood of obtaining an extreme sampling is very small. We conducted thousands of LHD samplings and did not see a single extreme case that was even close to the PS case. It was much more likely to obtain much better designs than the extreme designs. Much more and deeper study is needed to find a very reliable measure for absolute uniformity check.

7. CONCLUSIONS AND DISCUSSION

The effect of different design types based on different criteria or combinations thereof was studied. For the cases tested, the impact of design type on the prediction accuracy was modest. Among those cases that did exhibit significant differences, Latin hypercube design (LHD) was usually better than random design (Rd). When reporting on the relative performance of LHDs and RDs in other contexts, other authors also preferred LHDs, see for example McKay, Beckman, and Conover (1979), Owen (1992[b]), and Stein (1987). Maximin LHD (LHMm) was often worse than both LHD and Minimized CL2 Latin hypercube design (LHCL2) especially in 5D and 10D cases. LHMm was sometimes better in 2D cases when the sample size was close to 20. In these cases, the relative errors were often close to 0.01, 0.001 or even lower. LHCL2 was often better than LHD and LHMm in 2D cases. However, in 5D and 10D cases, LHD appeared to be better than the other designs, and LHCL2 was usually better than LHMm. Uniform designs based on the tables on the web were sometimes better than any of the other designs, but worse in many other cases.

Among the designs tested, LHCL2 or LHD is recommended for 2D cases; LHD is recommended for 5D and 10D cases, with less sampling effort and better prediction results. In any case, a check is needed to see if LHD produces a very nonuniform design. If so, another LHD must be generated.

Although the testing range was limited, this study revealed that it was more important to have enough samples than to have better designs in most cases, unless the design was extremely nonuniform. It was often necessary to use enough samples to reach desired error levels by one-stage sampling (vs. sequential sampling). The detailed guidelines for sample size determination may be very useful for the users of computer experiments. It is desirable that the knowledge of the tradeoff between the effort in sampling and modeling and the result (accuracy level) be available to the users.

It has been shown that, in general, more uniformly distributed sampling did not lead to more accurate modeling in computer experiments in 2D and 3D cases, unless the designs included extremely nonuniform cases. This result seems likely to be true for higher dimensions as well. Thus, it may not be appropriate to use uniformity as the sole criterion to compare different experimental designs or as the only goal to be pursued in the search for better designs, although it is true that very nonuniform samplings should be avoided. Much research effort has been devoted to finding "better" designs with more uniformly distributed sampling. It seems that other goals besides uniformity need to be studied. What is more important than uniformity or "better designs," at least when prediction accuracy is important, is to assure that sample size is large enough. Sample size determination and the development of good sequential design methods may warrant additional attention.

This study is a preliminary investigation for the topics studied. More comprehensive simulation tests are needed before more general conclusions can be drawn. In this study, twenty test functions were tested. Many more should be evaluated. The output from computer programs might behave differently from the test functions, thus it may be necessary to run tests using output from computer programs, for metamodeling purposes. The numbers of design variables tested were 2, 5, and 10. For large design problems, much larger numbers must be tested. Sample sizes up to 300 were tested. Much higher numbers may be needed for large problems. Only a portion of the design types in the literature was tested. The other types also need to be tested. Finally, since only kriging models were used in this study, other approximation models should be tested as well, such as support vector regression, genetic programming, neural networks, radial basis functions, multivariate adaptive regression splines (MARS), polynomials, etc.

For the effect of sampling uniformity, more and deeper study is needed beyond this preliminary investigation, including research into the definitions of uniformity (relative and absolute). Since sampling uniformity has been taken as a fundamental issue for computer experiments, the observations presented here may be useful to the users as well as researchers developing experimental designs.

For those designs with combined criteria for optimizing LHD, the fields of multi-objective optimization (MOO) or multidisciplinary optimization (MDO) may be instructive. The methodologies used in MOO or MDO might help to find lower combined criterion values or better designs. The use of MOO or MDO methodologies in that regard has not been tried. More study is needed to compare the designs based on combined criteria and to compare them with the designs based on a single criterion, within and outside the LHD domain.

We close this paper with a comment from a private communication from William Notz that we received on an earlier version of the manuscript in which he provided additional observations and a thoughtful analysis.

"The results you have found agree with what we have observed. Our experience is that any design that is reasonably uniform seems to work well. Only very nonuniform designs (for example, designs that take most of their observations on the boundary) seem to perform poorly. We believe the reason is that interesting features of response surfaces (locations of maxima, minima, regions where the response surface varies greatly) generated by simulations tend to be "uniformly" spread out over the design region. The variation in the location of these interesting features is larger than subtle variations in the uniformity of designs. Thus, only designs that take observations in a very limited portion of the design space perform poorly.

We also agree that the size of the sample is the most important feature, but this is not surprising. Only an extremely nonuniform design is likely to be outperformed by a uniform design of smaller sample size. In most of experimental design, sample size has a bigger impact than subtle differences in designs."

ACKNOWLEDGEMENT

Thanks go to Mr. Greg Saxton, PE, the Chief Engineer at Gunderson Inc., for his support and encouragement over many years. We gratefully acknowledge that the referee provided many insightful comments and criticisms. His/her comments resulted in many improvements including the point that a check is needed after a LHD is generated to guard against possibly very nonuniform designs. The results with proportional sampling were added to support the point and to show larger pictures, as he/she suggested. We are also grateful to Theodore Allen, Vladimir Balabanov, Selden Crary, Kai-Tai Fang, Anthony Giunta, William Notz, Brian Rutherford, Timothy Simpson, and Bo Ping Wang, who read the earlier manuscript and responded with similar observations, explanations, encouragement, and/or advice. We appreciate that James Gronseth, Pat Harmon, and Wayne Wakeland diligently worked through a later version to improve the writing. We also thank Multnomah County Library and the Portland State University Library for obtaining many research papers for us.

APPENDIX Test Functions

Note: x in $f(x)$ is a vector with coordinates $x_1,\ldots,x_n$.

1) Function 1 (AC): Ackley's path function

$$f(x) = -a * \exp(-b * \sqrt{(1/n)(\sum_{i=1}^{n} x_i^2)}) - \exp(\frac{1}{n}\sum_{i=1}^{n} \cos(cx_i)) + a + e$$

$a = 20$; $b = 0.2$; $c = 2 \cdot pi$; $I = 1:n$; $x_i \in [-30,30]$;

2) Function 2 (AX): Axis parallel hyper-ellipsoid function

$$f(x) = \sum_{i=1}^{n} ix_i^2$$

$x_i \in [-30,30]$;

3) Function 3 (DE): De Jong's function 1

$$f(x) = \sum_{i=1}^{n} x_i^2$$

$x_i \in [-30,30]$;

4) Function 4 (RB): Rosenbrock's valley (De Jong's function 2)

$$f(x) = \left[\sum_{i=1}^{n} 100(x_{i+1} - x_i^2)\right]^2 + (1 - x_i)^2$$

$x_i \in [-30,30]$;

5) Function 5 (RY): rotated hyper-ellipsoid function

$$f(x) = \sum_{i=1}^{n} \sum_{j=1}^{i} x_j^2$$

$x_i \in [-30,30]$;

6) Function 6 (MI): Michalewicz's function

$$f(x) = -\left(\sum_{i=1}^{n} \sin x_i\right) * \left[\frac{\sin(ix_i^2)}{\pi}\right]^{2m}$$

m=10;

$0 \le x(i) \le pi$ (π).

7) Function 7 (BR): Branins's rcos function

$$f(x_1, x_2) = a(x_2 - bx_1^2 + cx_1 - d)^2 + e(1 - f)\cos x_1 + e$$

a=1, b=5.1/(4·pi^2), c=5/pi, d=6, e=10, f=1/(8·pi);

$-5 \le x_1 \le 10, 0 \le x_2 \le 15$.

8) Function 8 (GD): Goldstein-Price's function

$$f(x_1, x_2) = [1 + (x_1 + x_2 + 1)^2 (19 - 14x_1 + 3x_1^2 - 14x_2 + 6x_1x_2 + 3x_2^2)] *$$
$$[30 + (2x_1 - 3x_2)^2 (18 - 32x_1 + 12x_1^2 + 48x_2 - 36x_1x_2 + 27x_2^2)]$$

$-2 \le x(i) \le 2$, i=1:2.

9) Function 9 (SX): Six-hump camel back function

$$f(x_1, x_2) = (4 - 2.1x_1^2 + \frac{x_1^4}{3})x_1^2 + x_1x_2 + (-4 + 4x_2^2)x_2^2$$

$-3 \le x_1 \le 3, -2 \le x_2 \le 2$.

10) Function 10 (PK): Peaks function

$$f(x_1, x_2) = 3(1 - x_1)^2 \exp(-x_1^2 - (x_2 + 1)^2) - 10(\frac{x_1}{5} - x_1^3 - x_2^5)\exp(-x_1^2 - x_2^2) -$$
$$\frac{1}{3}\exp(-(x_1 + 1)^2 - x_2^2)$$

$-3 \le x_1 \le 3, -3 \le x_2 \le 3.$

For functions 11-20: $x_i \in [-30, 30]$;

11) Froth function

$f(x_1, x_2) = -13 + x_1 + ((5 - x_2) x_2 - 2) x_2 - 29 + x_1 + ((x_2 + 1) x_2 - 14) x_2;$

12) Helix2 function

$$f(x_1, x_2) = 10(\sqrt{x_1^2 + x_2^2} - 1)$$

13) Rose function

$f(x_1, x_2) = 10(x_2 - x_1^2)$

14) Sing2 function

$f(x_1, x_2) = \sqrt{5}(x_1 - x_2);$

15) Sing3 function

$f(x_1, x_2) = (x_1 - 2 x_2)^2;$

16) Sing4 function

$f(x_1, x_2) = \sqrt{10}((x_1 - x_2)^2) ;$

17) Wood1 function

$f(x_1, x_2) = 10*(x_2 - x_1^2);$

18) Wood3 function

$f(x_1, x_2) = \sqrt{90}(x_2 - x_1^2);$

19) Wood5 function

$f(x_1, x_2) = \sqrt{10} (x_1 + x_2 - 2);$

20) Wood6 function

$f(x_1, x_2) = (x_1 - x_2)/\sqrt{10}$

a) p= 0.0102

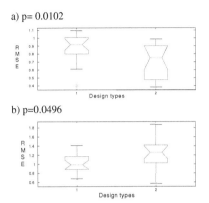

b) p=0.0496

Figure 1. ANOVA plots to show comparison with p-v close to 0.01 and 0.05.

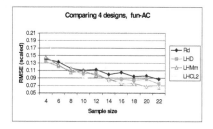

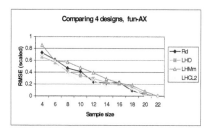

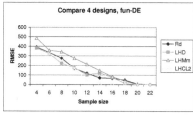

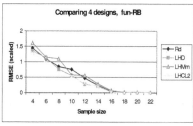

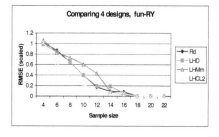

Figure 2 (a-e) RMSE vs. sample sizes, 2D, 4 designs, sample sizes from 4 to 22.

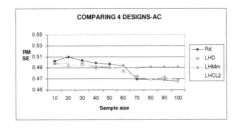

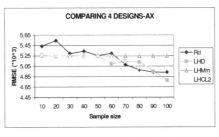

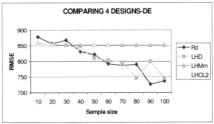

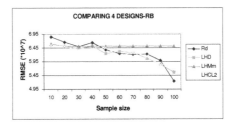

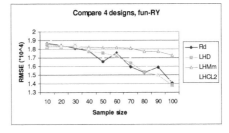

Figure 3 (a-e) RMSE vs. sample sizes, 10D, 4 designs, sample sizes from 10 to 100.

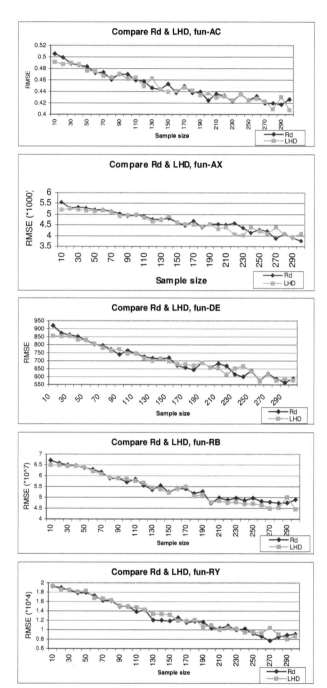

Figure 4 (a-e). Comparing Rd and LHD, n=10, m=10-300, functions AC,AX,DE,RB,RY.

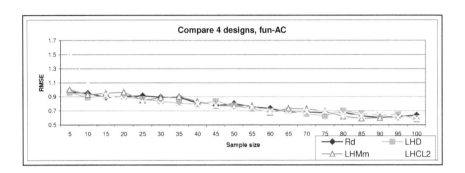

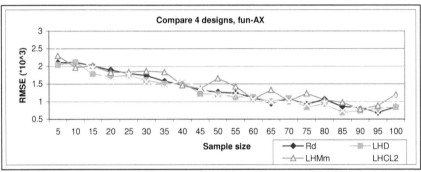

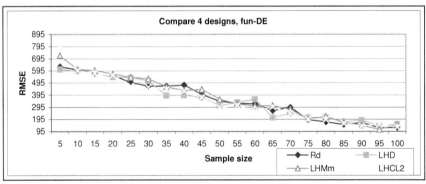

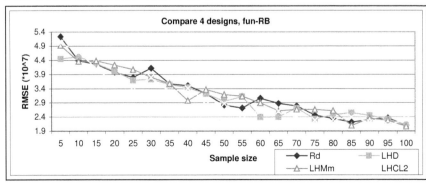

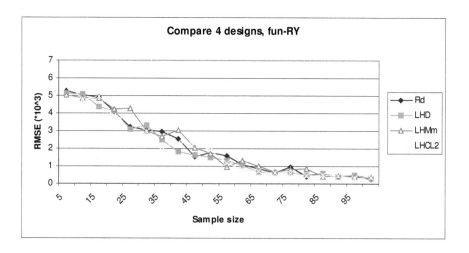

Figure 5. RMSE vs. sample sizes, 5D, 4 designs, sample sizes from 5 to 100. (function: AC,AX,DE,RB,

RY)

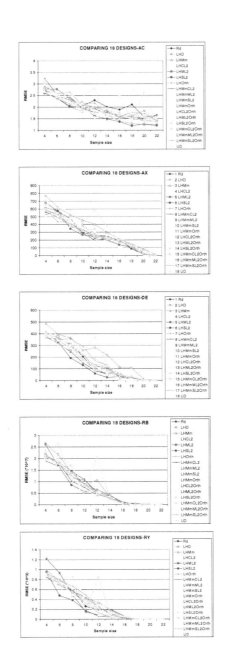

Figure 6 (a-e). RMSE vs. sample sizes, 2D, 18 designs, sample sizes from 2 to 22.

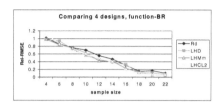

Figure 7. Comparing 4 designs for function BR

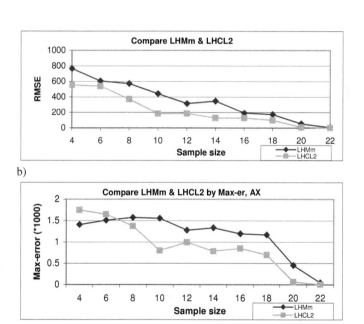

b)

Figure 8. Comparison between LHMm and LHCL2 based on RMSE and maximum error for function AX.

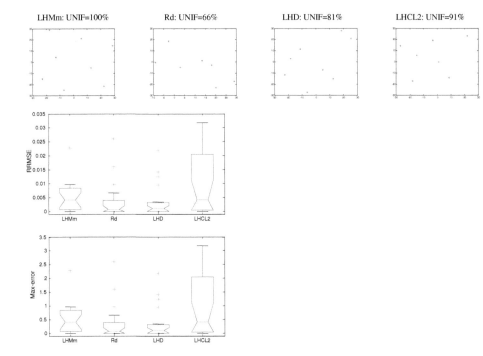

Figure 9. Comparing four designs based on comparing RRMSE, m=8 (m: sample size), p-v = 0.3975.

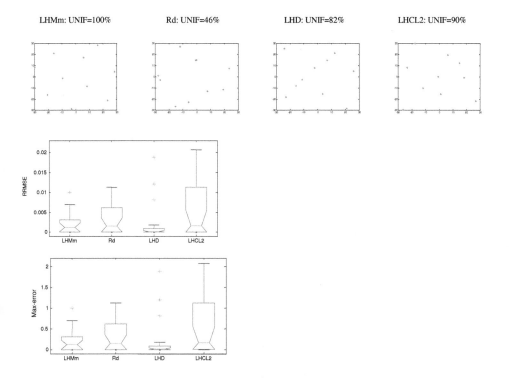

Figure 10. Comparing four designs based on comparing RRMSE, m=10 (m: sample size), p-v = 0.3974.

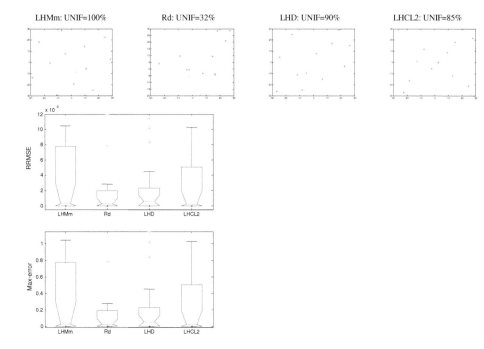

Figure 11. Comparing four designs based on comparing RRMSE, m=12 (m: sample size), p-v = 0.4104.

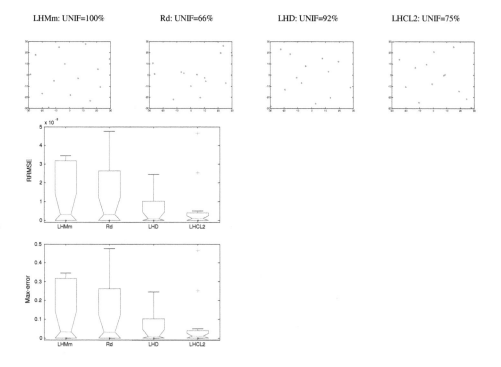

Figure 12. Comparing four designs based on comparing RRMSE, m=14 (m: sample size), p-v = 0.5340.

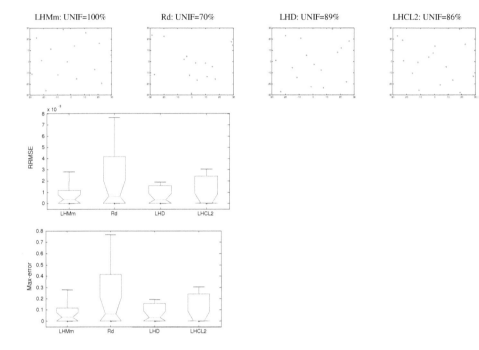

Figure 13. Comparing four designs based on comparing RRMSE, m=16 (m: sample size), p-v = 0.4634.

1) LHMm

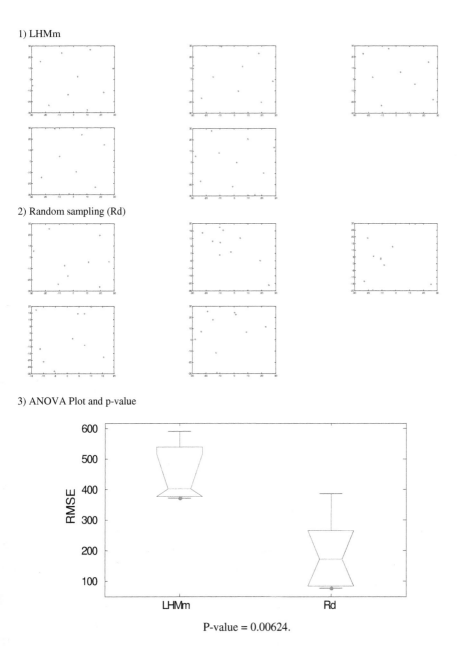

2) Random sampling (Rd)

3) ANOVA Plot and p-value

P-value = 0.00624.

Figure 14. Comparing LHMm and Rd based on comparing RMSE, m=10 (m: sample size), function= AC

XY Projection	XZ Projection	ZY Projection

LHMm: UNIF = 100%

Rd: UNIF = 45%

LHD: UNIF = 88%

LHCL2: UNIF = 82%

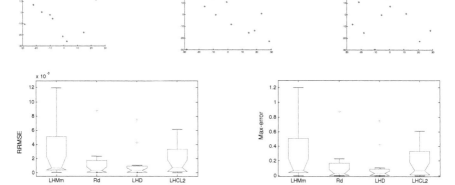

Figure 15. Comparing 4 designs by comparing
RRMSE and Max-error, 3D, m =12,
p-v = 0. 4427.

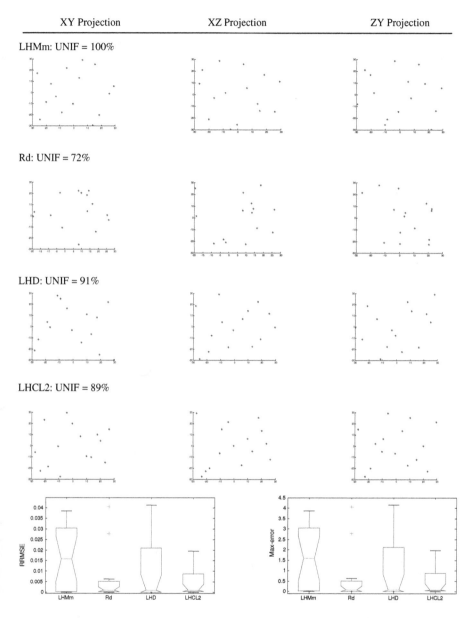

| XY Projection | XZ Projection | ZY Projection |

LHMm: UNIF = 100%

Rd: UNIF = 72%

LHD: UNIF = 91%

LHCL2: UNIF = 89%

Figure 16. Comparing 4 designs by comparing RRMSE and Max-error, 3D, m =15, p-v = 0. 0.6882.

XY Projection	XZ Projection	ZY Projection

LHMm: UNIF = 100%

Rd: UNIF = 73%

LHD: UNIF = 88%

LHCL2: UNIF = 90%

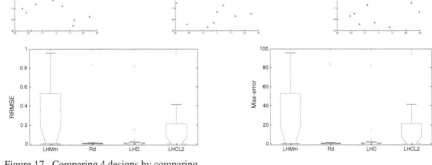

Figure 17. Comparing 4 designs by comparing RRMSE and Max-error, 3D, m =18, p-v = 0.5648.

XY Projection	Z Projection	ZY Projection

LHMm: UNIF = 100%

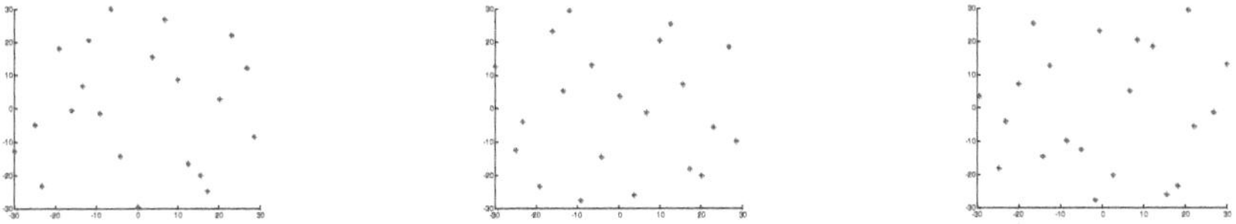

Rd: UNIF = 74

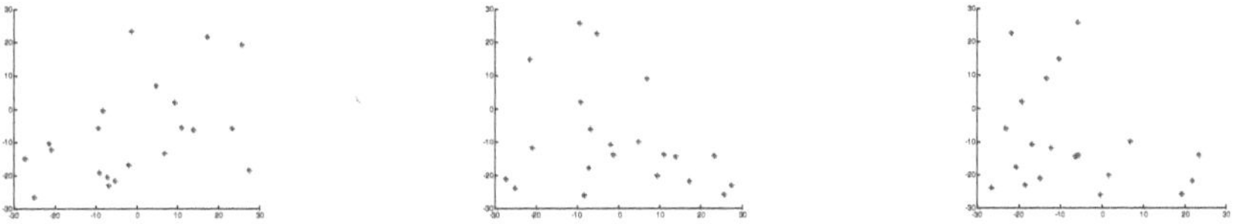

LHD: UNIF = 88%

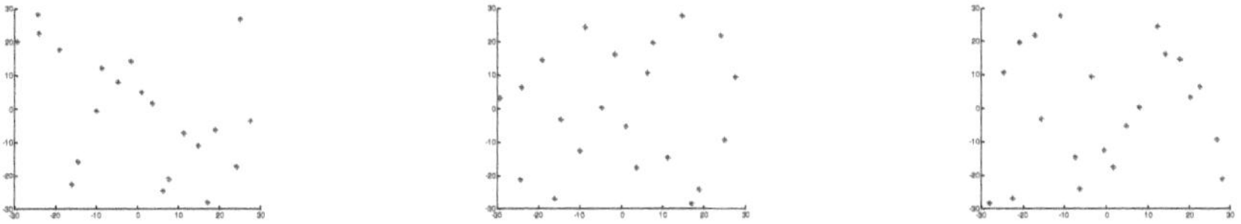

LHCL2: UNIF = 92%

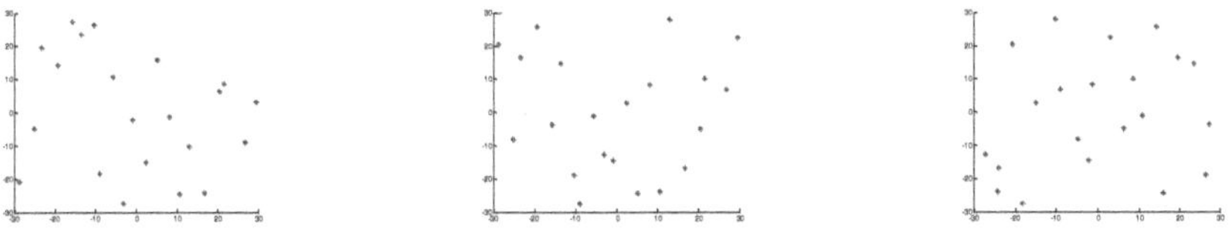

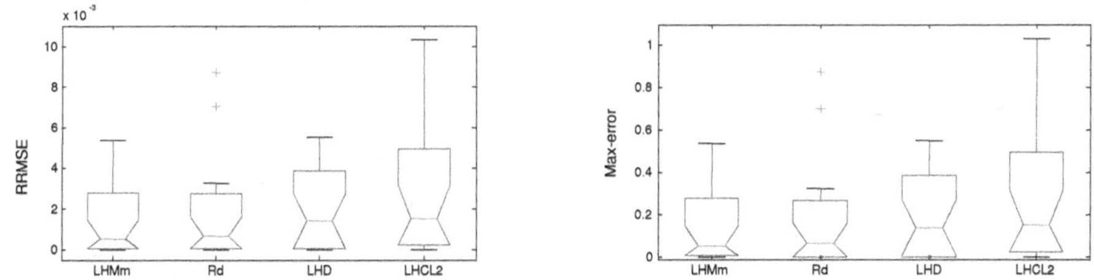

Figure 18. Comparing 4 designs by comparing RRMSE and Max-error, 3D, m =21, p-v = 0.8549.

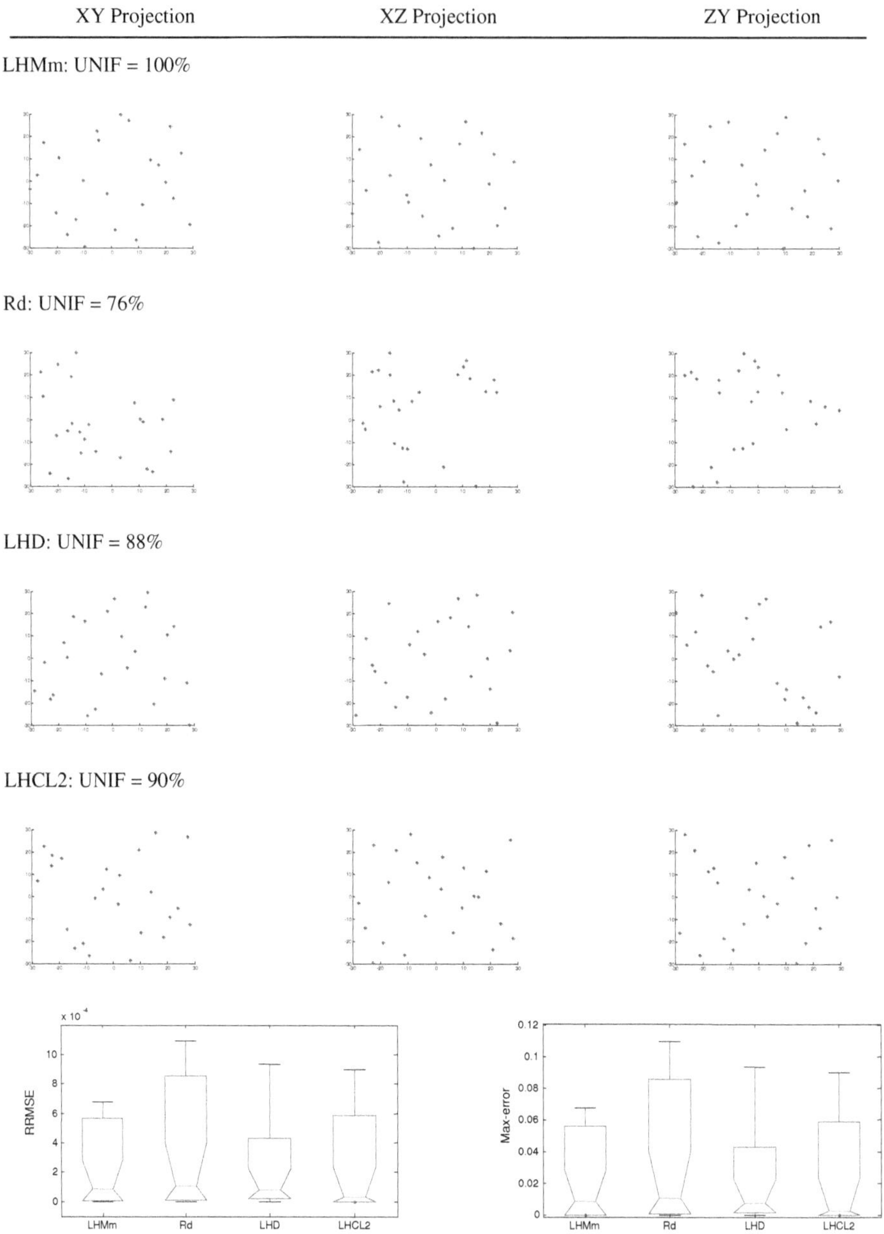
XY Projection
XZ Projection
ZY Projection
LHMm: UNIF = 100%
Rd: UNIF = 76%
LHD: UNIF = 88%
LHCL2: UNIF = 90%
x 10^-4
RRMSE
LHMm
Rd
LHD
LHCL2
Max-error
LHMm
Rd
LHD
LHCL2

Figure 19. Comparing 4 designs by comparing RRMSE and Max-error, 3D, m =24, p-v = 0.5313.

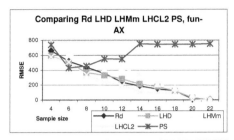

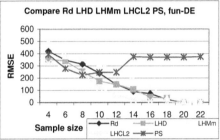

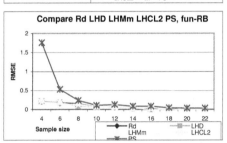

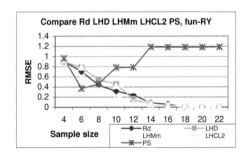

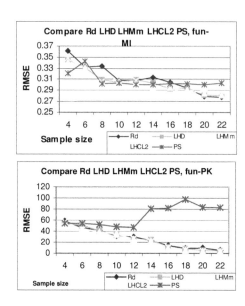

Figure 20 (a-e). Comparing Rd, LHD, LHMm, LHCL2, and PS; n=2, m=4-22, functions AX, DE, RB, RY, MI.

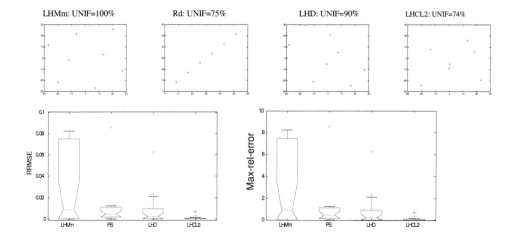

Figure 21. Comparing LHMm, PS, LHD, and LHCL2; 2D, m (sample size) =8, p-v = 0.2664

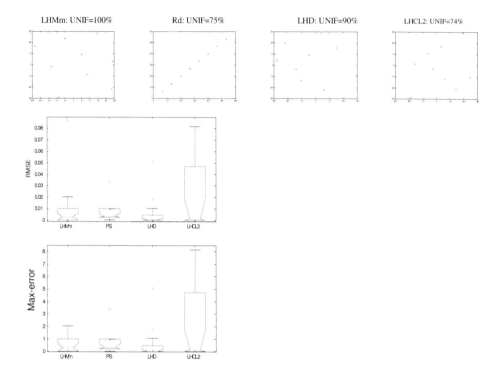

Figure 22. Comparing LHMm, PS, LHD, and LHCL2; 2D, m (sample size) = 10, p-v = 0.4887.

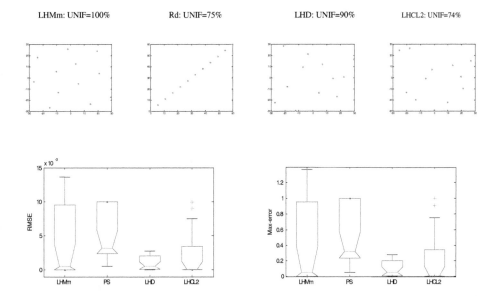

Figure 23. Comparing LHMm, PS, LHD, and LHCL2; 2D, m (sample size) = 12, p-v = 0.5653.

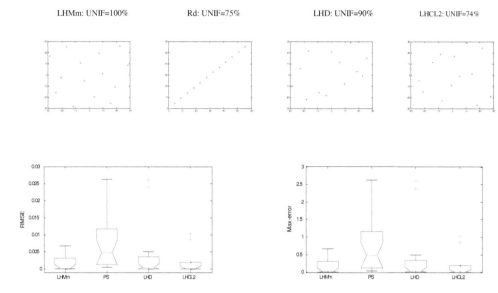

Figure 24. Comparing LHMm, PS, LHD, and LHCL2; 2D, m (sample size) = 14, p-v = 0.7135.

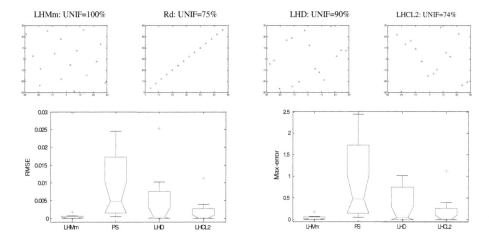

Figure 25. Comparing LHMm, PS, LHD, and LHCL2; 2D, m (sample size) = 14, p-v = 0.6474.

Table 1. Summary of the tests comparing Rd, LHD, LHMm, LHCL2

Comparing 4 design types Rd, LHD, LHMm, LHCL2			N = 2	N = 5	N = 10
			200 cases	50 cases	50 cases
			Percentage, %	Percentage, %	Percentage, %
1)	Compare RD LHD LHMm LHCL2				
	p-v < 0.01		43.5	14	26
	p-v < 0.05		54	24	42
2)	Compare Rd LHD				
	p-v < 0.01		8.5	2	4
		Rd better	0.5	0	0
		LHD better	8	2	4
	p-v < 0.05		24	2	14
		Rd better	1	0	2
		LHD better	23	2	12
3)	Compare LHD LHMm				
	p-v < 0.01		27	10	28
		LHD better	13.5	10	28
		LHMm better	13.5	0	0
	p-v < 0.05		42.5	22	48
		LHD better	25	22	46
		LHMm better	17.5	0	2
4)	Compare LHD LHCL2				
	p-v < 0.01		8.5	8	2
		LHD better	1	6	0
		LHCL2 better	7.5	2	2
	p-v < 0.05		15.5	20	12
		LHD better	1.5	12	10
		LHCL2 better	14	8	2
5)	Compare LHMm LHCL2				
	p-v < 0.01		39	12	30
		LHMm better	15	2	2
		LHCL2 better	24	10	28
	p-v < 0.05		51.5	28	46
		LHMm better	21.5	6	4
		LHCL2 better	30	22	42

Sample sizes: 4 to 22 for 2D, 5 to 50 for 5D, 10 to 100 for 10D); Design types compared: Rd (random sampling), LHD (Latin hypercube design), LHMm (Maximin LHD), and LHCL2 (Minimized center L_2 discrepancy LHD); N: dimensions; Percentage: the number of the cases where the different designs are not equivalent (at the given p-value) divided by the total number of cases in the group; p-v: probability value from ANOVA (one-way).

Table 2 Average RMSE values for comparing Rd, LHD, LHMm, and LHCL2 (function: BR)

m	Rd	LHD	LHMm	LHCL2
4	1.0125	0.9795	0.9833	1.1204
6	0.8597	0.9356	0.8427	0.8615
8	0.7604	0.7256	0.7576	0.6805
10	0.7009	0.5654	0.6325	0.5974
12	0.5543	0.4501	0.4216	0.4812
14	0.4662	0.3868	0.4177	0.3985
16	0.3103	0.3288	0.2288	0.1798
18	0.1514	0.1129	0.1218	0.1046
20	0.1654	0.1136	0.0622	0.0660
22	0.1040	0.0734	0.0527	0.0580

Table 3. Two-way ANOVA table for comparing 4 designs, function-BR

Source	SS	dof	MS	F
Columns (designs)	0.3062	3	0.1021	5.082
Rows (sample sizes)	61.4	9	6.822	339.7
Interaction	0.7511	27	0.02782	1.385
Error	11.25	560	0.02008	
Total	73.7	599		

Table 4. Variation analysis based on two-way ANOVA tables when comparing 4 designs

Functions	Design portion, %	Sample size portion, %	Others, %	Total, %
AC, 2D	2.5	48.0	49.5	100
AX, 2D	2.4	75.0	22.6	100
DE, 2D	3.3	75.9	20.8	100
RB, 2D	1.2	80.8	18.0	100
RY, 2D	1.5	78.7	19.8	100
MI, 2D	0.5	37.4	62.1	100
BR, 2D	0.4	83.4	16.2	100
GD, 2D	1.8	40.4	57.8	100
SX, 2D	6.8	50.0	43.2	100
PK, 2D	1.5	48.0	50.5	100
AC, 5D	1.1	30.9	68.0	100
AX, 5D	2.1	47.2	50.7	100
DE, 5D	1.3	61.6	37.1	100
RB, 5D	0.8	52.6	46.6	100
RY, 5D	0.6	67.0	32.4	100
AC, 10D	2.6	18.7	78.7	100
AX, 10D	5.9	12.4	81.7	100
DE, 10D	10.9	27.1	62.0	100
RB, 10D	8.4	23.3	68.3	100
RY, 10D	4.3	35.0	60.7	100

Percentage variation by designs, sample sizes, interaction plus within group variation (others), and the total (100%).

2D: m= 4, 6, …, 22; 5D: m= 5,10..,50; 10D, m= 10,…,100. m: sample size.

Table 5. FIINV values for samplings in Figures 21 – 25 (m: sample size)

m	LHMm	PS	LHD	LHCL2
8	1.1738	0.8820	1.0551	0.8632
10	0.6743	0.4888	0.6485	0.6011
12	0.4348	0.3056	0.4113	0.3934
14	0.3133	0.2071	0.2761	0.2904
16	0.2298	0.1485	0.1878	0.1886

Table 6. SFIINV values for samplings in Figures 21 – 25 (m: sample size)

m	LHMm	PS	LHD	LHCL2
8	212.5	159.7	191.0	156.3
10	213.2	154.6	205.1	190.1
12	216.9	152.4	205.2	196.2
14	229.8	151.9	202.5	213.0
16	235.3	152.1	192.3	193.1

References

Barton, R. R. (1994). December 11-14. Metamodeling: A State of the Art Review. Proceedings of the 1994 Winter Simulation Conference (Tew, J. D., Manivannan, S., et al., eds.), Lake Beuna Vista, FL, IEEE. 237-244.

Barton, R. R. (1998). December 13-16. Simulation Metamodels. Proceedings of the 1998 Winter Simulation Conference (WSC'98) (Medeiros, D. J., Watson, E. F., et al., eds.), Washington, DC, IEEE. 167-174.

Barton, R. R. (1992). Metamodels for simulation input- output relations. Proceedings of the 1992 Winter Simulation Conference, ed. D. Goldsman, J. Swain, J. Wilson, IEEE, Piscataway, New Jersey. 41 1 - 41 8.

Chipperfield, A., Fleming, P., Pohlheim, H., and Fonseca, C. *Genetic Algorithm Toolbox: User's guide*. http://www.shef.ac.uk/~gaipp/ga-toolbox/manual.pdf.

Fang, K. T. (1980). Experimental Design By Uniform Distribution. Acta Mathematice Applicatae Sinic, 3: 363-372.

Fang, K.-T. and Wang, Y. (1994). *Number-theoretic Methods in Statistics*. New York: Chapman & Hall.

Fang, K.T., Lin, D.K.J., Winker, P. and Zhang, Y. (2000). Uniform design: Theory and applications. Technometrics. 42:237-248.

Fang, K.T., Lin, D., and Liu, M. (2000). Optimal mixed-level supersaturated design and computer experiment. http://citeseer.ist.psu.edu/cache/papers/cs/17273/ftp.

Fang, K.T. , Ma, C.X. and Winker, P. (2002). Centered L2-discrepancy of random sampling and Latin hypercube design, and construction of uniform designs. Mathematics of Computation. 71:275-296.

Goldberg D.E. (1989). *Genetic algorithms in search, optimization and machine learning.* Reading, Massachusetts: Addison-Wesley Publishing Co., Inc.,

Hickernell, F.J. (1998). A generalized discrepancy and quadrature error bound. Mathematics of Computation. 67:299-322.

Holland J. (1975). *Adaptation in natural and artificial systems.* Ann Arbor: University of Michigan Press.

Iman, R.L. and Conover, W.J. (1982). A distribution-free approach to inducing rank correlation among input variables. Communications in Statistics. Series B 11:311-334.

Jin, R., Chen W., and Sudjianto, A. (2003). Sept. 2-6. An efficient algorithm for constructing optimal design of computer experiments. ASME 2003 Design engineering technical conferences, Chicago, Illinois, USA.

Johnson, M. E., Moore, L. M. and Ylvisaker, D. (1990). Minimax and Maximin Distance Designs. Journal of Statistical Planning and Inference. 26: 131-148.

Koehler, J. R., Owen, A. B. 1996: Computer experiments. In Ghosh, S. and Rao, C. R., editors, *Handbook of Statistics*, **13**:261--308. New York: Elsevier Science.

Liefvendahl, M. and Stocki, R: A study on algorithms for optimization of Latin hypercubes. http://www.ippt.gov.pl/~rstocki/JSPI-olh.pdf.

Li, W., and Wu, C.F.J. (1997). Columnwise-Pairwise algorithms with applications to the construction of supersaturated designs. Technometrics. 39:171-179.

McKay, M.D., Beckman R.J. and Conover W.J. (1979). A comparison of three methods for selecting values of input variables in the analysis of output from a computer code. Technometrics. **21**:239-245. (reprinted in 2000: *Technometrics*. **42**:55-61).

Morris, M. D. and Mitchell, T. J. (1995). Exploratory Designs for Computer Experiments. Journal of Statistical Planning and Inference. 43:381-402.

Nielsen, H.B., Lophaven, S.N., and Søndergaard, J.: DACE: A MATLAB Kriging Toolbox, http://www.imm.dtu.dk/~hbn/dace/.

Osio, I. G. and Amon, C. H. (1996) An Engineering Design Methodology with Multistage Bayesian Surrogates and Optimal Sampling. Research in Engineering Design. 8: 189-206.

Owen, A. B. (1992). Orthogonal Arrays for Computer Experiments, Integration and Visualization. Statistica Sinica, 2: 439-452.

Owen, A. B. (1992[b]). A Central Limit Theorem for Latin Hypercube Sampling. Journal of the Royal Statistical Society Ser. B 54:541-551.

Owen, A.B. (1994). Controlling correlations in latin hypercube samples. Journal of the American Statistical Association. 89:1517-1522.

Palmer, K. et al. (2001). Minimum bias Latin hypercube design. IIE Transactions. 33 :793-808.

Park, J.-S. (1994). Optimal Latin-Hypercube Designs for Computer Experiments. Journal of Statistical Planning and Inference. 39: 95-111.

Sacks, J., Welch, W.J., Mitchell, T.J. and Wynn, H.P. (1989). Design and Analysis of Computer Experiments. Statistical Science 4(4): 409-435.

Sacks, J., Schiller, S.B. and Welch, W.J. (1989). Designs for Computer Experiments. Technometrics. 31:41-47.

Sahama, T. and Diamond, N.T. (2001). Sample size considerations and augmentation of computer experiments. Journal of Statistical computation and simulation. 68:307-319.

Santner, T., Williams, B, Notz, W. (2003). *Design and Analysis of Computer Experiments.* New York: Springer-Verlag,

Shewry, M.C. and Wynn, H.P. (1987). Maximum entropy sampling. Journal of Applied Statistics. 14:165-170.

Simpson, T. W., Mauery, T. M., Korte, J. J. and Mistree, F. (2001). Kriging Metamodels for Global Approximation in Simulation-Based Multidisciplinary Design Optimization. AIAA Journal. 39: 2233-2241.

Simpson, T. W., Booker, A. J., Ghosh, D., Giunta, A. A., Koch, P. N., and Yang, R.-J. (2000). Approximation Methods in Multidisciplinary Analysis and Optimization: A Panel Discussion. Structural and Multidisciplinary Optimization. 20:124-137.

Simpson, T. W., Peplinski, J., Koch, P. N. and Allen, J. K. (2001). Metamodels for Computer-Based Engineering Design: Survey and Recommendations. Engineering with Computers. 17: 129-150.

Stein, M. (1987). Large Sample Properties of Simulations Using Latin Hypercube Sampling. Technometrics. 29:143-151.

Tang, B., (1993). Orthogonal Array-Based Latin Hypercubes. Journal of the American Statistical Association. 88: 1392-1397.

Tang, B. (1994). A Theorem for Selecting OA-Based Latin Hypercubes Using a Distance Criterion. Communications in Statistics, Theory and Methods. 23: 2047-2058.

Welch, W. J., Buck, R. J., Sacks, J., Wynn, H. P., Mitchell, T. J. and Morris, M. D. (1992). Screening, Predicting, and Computer Experiments. Technometrics. 34: 15-25.

Welch, W. J., Yu, T.-K., Kang, S. M. and Sacks, J.. (1990). Computer Experiments for Quality Control by Parameter Design. Journal of Quality Technolog. 22: 15-22.